Current Natural Sciences

Vincent BALTZ

The Basics of Electron Transport in Spintronics

Textbook with Lectures, Exercises and Solutions

Science Press
edp sciences

Printed in France

EDP Sciences – ISBN(print): 978-2-7598-2917-0 – ISBN(ebook): 978-2-7598-2918-7
DOI: 10.1051/978-2-7598-2917-0

Foreword

It is almost 200 years, since Georg Ohm published his research on the galvanic response of a metal to an applied voltage, amid a storm of controversy. He gave us, one of the handfuls of physics equations that everybody learns, $V = IR$. Now, we prefer to relate the linear galvanic response to the electric stimulus, via the intrinsic electrical conductivity, σ, of the metal, $\boldsymbol{j} = \sigma \boldsymbol{E}$. Some 20 years later, William Thomson discovered both the quadratic decrease of conductivity of a normal metal in a magnetic field – the normal magnetoresistance – and its variation with the angle between the electric current and the magnetization $\boldsymbol{M}$, in a ferromagnetic metal. Near the end of the 19th century, Edmund Hall found that in a perpendicular magnetic field, a current excites a transverse voltage proportional to the current in a normal metal, and proportional to magnetization in a ferromagnet – the normal and anomalous Hall effects.

Shortly afterwards, the electron with its tiny mass and unchanging quantum of negative charge $-e$ was identified as the mobile carrier of electric current, but its ability to transport the angular momentum associated with its intrinsic spin of $\hbar/2$ discovered in the 1920s, attracted little attention. This was because, unlike the charge, the angular momentum of the electron could be flipped in a collision with another electron. Spin diffusion lengths, measured in nanometres, are a small multiple of the electrons' mean free path. Anisotropic magnetoresistance and anomalous Hall effect became familiar effects but remained unexplained for much of the 20th century. Only when it became possible, in the 1970s to prepare high-quality magnetic thin films and heterostructures, thinner than the spin diffusion length, did 'spin electronics' become a practical possibility.

Since then, there has been an avalanche of discovery of new magneto-electric phenomena, and spintronics has found important applications in contactless sensing, scalable non-volatile memory, and fast electronic switching. A bewildering array of new ideas and phenomena has emerged, many associated with spin-orbit interaction. Topology in direct and reciprocal space is an important consideration.

DOI: 10.1051/978-2-7598-2917-0.c901

A page is needed to list the acronyms, let alone explain the physical effects. There is much for a newcomer to the field to master, both conceptually and practically. Vincent Baltz's new book is a welcome guide to the first aspect, for both newcomers and practitioners of the art of spin-dependent electron transport.

Here is a concise, meticulously-illustrated account of the subject with necessary, but not excessive mathematical detail, which will allow the reader to grasp the basic concepts and learn how to use them in the ten extended exercises, that form an integral part of the text. A common notation and a single system of practical SI units is adopted throughout, which helps to reduce the confusion of conventions and units, found in the literature. Familiarity with the numerical values of the quantities involved will allow the reader to develop a critical, physical feel for the subject. Baltz's book is based on a series of lectures, given to students in Grenoble, in recent years. It will serve as a Baedeker for travellers in the rugged and testing terrain of contemporary spintronics.

Michael COEY
Dublin, November 2022

Preface

This book is intended for readers, who wish to acquire a solid basis for understanding electron transport in spintronics and the fundamental principles of some associated applications. It provides the reader with sufficient knowledge, to be able to invest further in this field of research. To this end, some of the prominent key notions widely used in the field were selected with care, with the aim of providing a simple, concise, and efficient framework. These selected notions are explained, using simple examples and analytical calculations. The technical terms and specialist jargon are explained, and the way subtleties complicate the phenomena, without altering their physical basis, is addressed.

I am grateful to Mike Coey of Trinity College Dublin, and Daria Gusakova, Gilles Gaudin, and Matthieu Jamet of SPINTEC Grenoble, who took the time to read this book and gave me to benefit from their opinions, advice, and to correct some of the errors. I am very thankful to many students, as well as to colleagues from SPINTEC, and within the scientific community, for numerous stimulating discussions and for motivating my curiosity and interest in this topic. Finally, I would like to thank Hélène Béa and Olivier Fruchart, for allowing me to teach spintronics to complete their courses on nanomagnetism, at the University of Grenoble Alpes.

The text has grown out of lectures, given to MSc and PhD students. Therefore, it is designed for advanced graduates and postgraduates, as well as researchers and engineers, with a background in condensed matter physics and magnetism. Interested readers are encouraged to complement their knowledge by consulting, for example, the following books: N. W. Ashcroft and N. D. Mermin, *Solid State Physics*, Saunders College Philadelphia (1976); C. Kittel, *Introduction to Solid State Physics*, John Wiley & Sons (2004); J. M. D. Coey, *Magnetism and Magnetic Materials*, Cambridge University Press (2010); E. du Tremolet de Lacheisserie *et al.* (eds) *Magnetism I & II*, EDP Sciences (2002).

DOI: 10.1051/978-2-7598-2917-0.c902

The text is structured in seven parts. Each part feeds on the previous one. A careful effort has been made, to indicate how the parts and the physical phenomena, they describe, are related to each other. While the text focuses on the fundamental aspects, the implementation of the fundamental physical effects in typical applications is described in detail. Orders of magnitude of the various parameters and key phenomena are provided in each section, based on experimental data. The references are concise, in order to provide a reasonable initial framework for the interested reader, to explore further. Physical effects, that are more peripheral to current concerns are mentioned wherever relevant, with references. An Index of key concepts allows the reader to navigate from one key area to another. Throughout this textbook, we use SI units. A List of symbols and units, as well as the corresponding formulas, is provided so that the reader can refer to them throughout the reading. When there are several possible definitions in the literature, the definition used in this book is made explicit, and the differences in prefactor depending on the articles and books consulted, are also made explicit.

Chapter 1 gives a concise overview of spintronics and what this discipline contributes to society.

Chapter 2 lays the foundation of spin-dependent electron transport. Based on the two-current and *sd* scattering models, we first show why the electron scattering probability is spin-dependent, and how this effect impacts electron transport, using the example of the current-in-plane (CIP) giant magnetoresistance (GMR) effect. We then show how the addition of spin-orbit interactions mixes the spin states and reshuffles the spin-dependent scattering probabilities, using the anisotropic magnetoresistance (AMR) effect as an example.

In chapter 3, we describe in detail, the effect of spin accumulation, encountered whenever electrons flow across an interface, due to the distinct partial current densities in materials of different types. Here, we detail how the electron dynamics are described in this case, and what kind of spin-flip relaxation mechanisms prevail. We elaborate on the conditions of a key parameter known as the spin-coupled interface resistance, necessary to maintain the spin polarization across an interface, and present how spin accumulation is at the heart of the current perpendicular-to-plane (CPP) GMR effect. How intrinsic interface effects, such as spin memory loss, as well as how non-collinearity and non-uniformity alter spin accumulation, are presented.

Chapter 4 focuses on the process by which spin angular momentum can be transferred from current (spin angular momentum flow) to magnetization, and the type of torque (STT), that this transfer generates. The key parameters involved, including the spin mixing conductance, are discussed in detail. How STT alters electron transport, by providing an additional relaxation channel for the spins, is tackled, as well as how STT can trigger oscillations and magnetization reversal. The spin pumping (SP) reciprocal effect of STT is introduced. Guidance on the morphology of STT in non-uniform magnetic textures, as well as hints on the magnetoelectronic circuit theory widely used in current spintronics, will be given. This chapter ends with a section about spin-orbit torques (SOT), which result from the transfer of angular momentum between lattice and the electron orbitals (crystal field potential), electron orbital and spin (spin-orbit coupling), and spin and magnetization (*sd*-exchange interactions).

In chapter 5, we describe how the intrinsic crystal and spin symmetries, represented by different symmetry groups, are particularly important notions, that have a significant impact on spintronics. With the examples of the different Hall effects, be they unquantized or quantized, we present the Berry formalism, and what it implies for intrinsic physical effects. The importance of breaking spatial or temporal inversion symmetries is detailed and illustrated.

Chapter 6 provides a series of ten comprehensive exercises with solutions. They are specifically designed to illustrate the ideas in the previous chapters and lead to other spintronic effects based on these ideas. The exercises deal with AMR (chapter 2), domain wall AMR (chapter 3), the drift-diffusion equation for spin accumulation (chapters 3–5), spin conductivity mismatch in CPP-GMR (chapter 3), the intrinsic intra-band scattering contribution to damping of magnetization dynamics (chapters 2 and 4), spin pumping and the inverse spin Hall effects (ISHE) (chapters 4 and 5), the extrinsic spin pumping contribution to damping (chapter 6), spin Hall magnetoresistance (SMR) (chapters 2–4), the harmonic analysis of the anomalous Hall voltage and related torques (chapters 4 and 5), and the intrinsic anomalous Hall and Nernst effects (AHE, ANE) (chapter 5).

Finally, chapter 7 concludes the book with a non-exhaustive presentation of some current topics, to which the readers may wish to turn, building on the previous chapters, which now allow them to go further, and understand other related or more elaborate spintronic phenomena.

Vincent BALTZ
Grenoble, November 2022

Contents

Chapter 1

Introduction

Spintronics consists of the investigation and exploitation of the electron spin degree of freedom in electronics. The multidisciplinary nature of spintronics requires familiarity with general aspects of condensed matter, notably in nanomagnetism [J. M. D. Coey, *Magnetism and Magnetic Materials*, Cambridge University Press (2010); E. du Tremolet de Lacheisserie *et al.* (eds) *Magnetism I & II*, EDP Sciences (2002)] and spin transport, which are complementary disciplines.

Spin $\boldsymbol{s}$ is one of the internal quantum properties of the electron, just like mass m_e or electric charge $-e$. It has no classical equivalent, but, by analogy with orbital angular momentum, it is often equated with classical angular momentum $(m_e\boldsymbol{r} \times \boldsymbol{v})$, based on the rotational movement of a uniform mass with velocity $\boldsymbol{v}$ on a circle of radius $\boldsymbol{r}$ (figure 1). The unit of spin is, therefore, the same as classical angular momentum, N.m.s^{-1} or J.s. The spin of an electron is quantized and can take two values $\boldsymbol{s} = \pm\hbar/2$ where $\hbar$ is Planck constant h divided by 2π. These two values are related to the spin projection quantum number, $m_s = \pm 1/2$, with $\boldsymbol{s} = \pm m_s\hbar$, for a spin quantum number $s = 1/2$ and a corresponding amplitude, $\mathcal{S} = \hbar\sqrt{s(s+1)}$.

Note: throughout the text, bold is used for vectors, $\boldsymbol{r} = \overrightarrow{r}$. All symbols and units are listed on page 150.

Because an electron has a spin and a charge, it also has a magnetic moment $\boldsymbol{m}$, just like a rotating electrically charged body in classical electrodynamics, which gives rise to a magnetic dipole moment, $(\propto -\frac{e}{2}\boldsymbol{r} \times \boldsymbol{v})$ (figure 1).

The actual relation between the spin and the magnetic moment of an electron is derived from quantum mechanics

$$\boldsymbol{m} = -\frac{e}{2m_e} g\boldsymbol{s} \tag{1}$$

Compared to classical mechanics, the result of quantum mechanics involves the Landé g-factor due to relativistic effects. For an isolated electron, $g \sim 2$. An electron thus takes values $\boldsymbol{m} \sim \mp \mu_{\mathrm{B}}$, with $\mu_{\mathrm{B}} = \frac{e\hbar}{2m_e}$ the Bohr magneton, in A.m^2.

DOI: 10.1051/978-2-7598-2917-0.c001

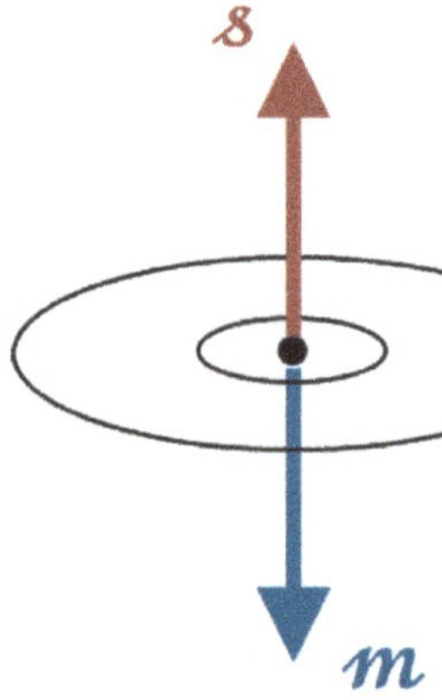

FIG. 1 – Illustration of the electron spin $\mathcal{s}$ and magnetic moment $\mathcal{m}$, associated with the rotation of a uniform distribution of charge over mass $\frac{e}{m_e}$, represented by the arrow in the horizontal plane.

The expression above is often used as,

$$\mathcal{m} = \gamma \mathcal{s} = -|\gamma| \mathcal{s} \tag{2}$$

where $\gamma = -\frac{e}{2m_e} g < 0$ is the gyromagnetic ratio. For an isolated electron, $\frac{|\gamma|}{2\pi} \sim 28\,\text{GHz.T}^{-1}$.

In matter, the total spin angular momentum takes into account the spin and orbital angular momenta, as well as the coupling between the two, called spin-orbit coupling, and possibly, some quenching of the orbital angular momentum due to crystal symmetry. The resulting total spin angular momentum and the corresponding magnetic moment take on discrete values, the amplitudes and projections of which are also defined by a principal j and a projection or secondary m_j quantum number. The relation between the total spin angular momentum and the total magnetic moment takes the same form as equation (1), except that the value of g now takes into account spin and orbital contributions, and possibly, spin-orbit coupling and some orbital quenching, for example, $g_{\text{Fe}} \sim 2.09$, $g_{\text{Ni}} \sim 2.19$, and $g_{\text{Co}} \sim 2.15$.

It is important to note that, in the literature, the term spin is sometimes used interchangeably for spin magnetic moment $\mathcal{m}$ and spin angular momentum $\mathcal{s}$, which can lead to confusion as $\mathcal{m}$ and $\mathcal{s}$ are of opposite sign due to the negative charge $-e$ of the electron. **In this book, the term moment will refer exclusively to $\mathcal{m}$ and spin to $\mathcal{s}$.** This distinction is crucial when dealing with the accumulation or transfer of angular momentum, particularly with regard to the sign of the coupling between several angular momenta, at the interface between two materials.

Spin transport, especially when conduction electrons carry out spin, is the subject of this book. In fact, spin information can be conveyed in two direct ways (figure 2), *via:*

- **electrons**, *i.e.*, *via* the spin of the conduction electrons, thus complementing the charge transport – although the net charge transport may be zero in some cases (chapter 2),
- **magnons**, *i.e.*, *via* the exchange interactions between the spin of the localized electrons in the crystal lattice, giving rise to so-called spin waves, quantified by elementary excitations called magnons. Magnon-mediated transport may be considered to be a separate subject, beyond the scope of this book. The interested reader can find more information in S. Rezende, *Fundamentals of Magnonics*, Springer (2020).

We note that starting from conduction electrons and magnons, there are several indirect ways to convert and transport spin. More generally, spin is a form of angular momentum tied to electrons but any excitation that possesses angular momentum (chapter 2), like phonons and photons, has the ability to be a carrier. Any angular momentum transfer process will be allowed *via* some interconversion, *e.g.*, electron–electron (chapters 2–4), electron–magnon, electron–phonon, and electron–photon.

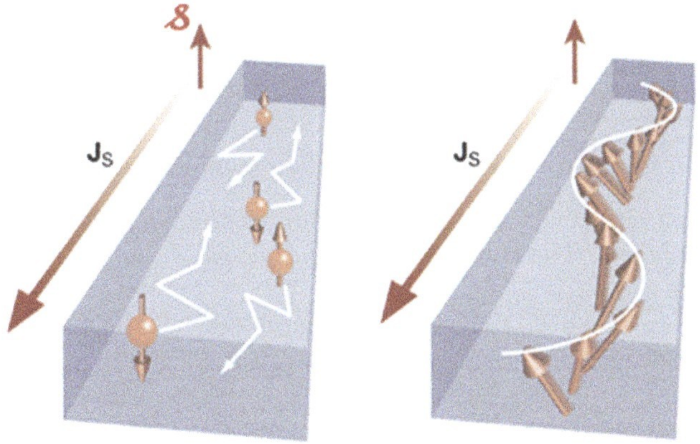

FIG. 2 – Illustration of how spin information can be conveyed, by (Left) itinerant conduction electrons and (Right) collective excitations of the localized moments in the crystal lattice. $\boldsymbol{J}_s$ represents the spin current density, and the arrow pointing up indicates the spin angular momentum $\mathcal{S}$, which gives the polarization of $\boldsymbol{J}_s$ (see chapter 2). Note: in the illustrations, the net charge current is zero, which is a special case detailed in chapter 2 for itinerant conduction electrons. Reprinted with permission from Springer Nature: Y. Kajiwara *et al.*, *Nature* **464**, 262 (2010). Copyright 2010.

The field of spintronics builds on several success stories that have contributed to technology and to society. Two of them trace the scientific path from two fundamental discoveries to industrial applications, involving magnetic sensors that are sold in the hundreds of billions:

- **from the giant magnetoresistance (GMR) effect to sensors for automotive and IT applications**

  The understanding of the basics of electron transport in relation to the GMR effect and the use of this fundamental effect in some applications will be discussed in detail in chapters 2 and 3. The chronology of the success of GMR is:

  1988 Discovery of the GMR effect [M. N. Baibich *et al.*, *Phys. Rev. Lett.* **61**, 2472 (1988); G. Binasch *et al.*, *Phys. Rev. B* **39**, 4828 (1989)]

  1991 Development and practical implementation in spin-valves [B. Dieny *et al.*, *J. Magn. Magn. Mater.* **93**, 101 (1991); S. S. P. Parkin, *Phys. Rev. Lett.* **64**, 2304 (1990)]

  1997-present Commercial product – sensors for automotive and information technology (IT) as billions of hard-disk drives – since supplanted by tunnel magnetoresistance (TMR) junctions in some applications [demonstration of TMR: J. S. Moodera *et al.*, *Phys. Rev. Lett.* **74**, 3273 (1995); the proposition of orbital-filtering for larger TMR values: W. H. Butler *et al.*, *Phys. Rev. B.* **63**, 054416 (2001)]

  2007 Nobel prize in physics awarded to A. Fert and P. Grünberg [https://www.nobelprize.org/]

- **from the spin transfer torque (STT) effect to memories for IT applications**

  The basics of electron transport in relation to the STT effect are discussed in detail in chapter 4. The chronology of the success of the STT effect is:

  1995 Discovery of the STT effect [J. C. Slonczewski, *J. Magn. Magn. Mater.* **159**, L1 (1996), L. Berger, *Phys. Rev. B* **54**, 9353 (1996)]

  2005 Practical implementation in TMR junctions

  2013-present Commercial product – MRAM (magnetic random-access memory) [B. Dieny *et al.* (eds), *Introduction to Magnetic Random-Access Memory*, Wiley-IEEE Press (2016)]

There are still more successes to come, not least because of the emergence of new fundamental phenomena, such as pure spin current physics (chapter 2), spin-orbit torques (chapter 4) and Berry curvature physics (chapter 5), and the commercialisation of new MRAM devices of several types, including SOT-MRAMs by major microelectronics companies. The miniaturisation of electronics is hampered, in particular, by problems of power consumption. Microprocessor chips require about 100 $W.cm^{-2}$ to operate, which is an order of magnitude higher than a kitchen stove. In addition, energy consumption in a chip is aggravated by leakage [C. Singh and R. Tangirala, *IBS report* (2013)] and by the volatile nature of the data storage.

Spintronics is ideally positioned to circumvent energy consumption issues, as it can combine non-volatility with dissipation-free transport of spin information when it uses pure spin currents. Energy harvesting spintronics is predicted to be a viable potential technology for many application areas [A. Hirohata *et al.*, *J. Magn. Magn. Mater.* **509**, 166711 (2020)], for example:

- **Information technology** – **IT**, *e.g.*, memory, processors, data security
- **Biomedicine,** *e.g.*, sensors
- **Telecommunication**, *e.g.*, transceiver
- **Artificial intelligence** – **AI**, *e.g.*, neuromorphic computing

This wealth of applications of interest to society promises a bright future for the discipline, whether approached from the point of view of fundamental physics (see chapter 7 for a description of some of the current fundamental topics) or from an applications perspective. To further emphasise how magnetism and spintronics are and will be present in our lives, I would quote M. Coey: "More transistors and magnets are produced in fabs than grains of rice are grown in paddy fields".

It seems important to emphasize that fundamental research and the industrial developments that enable its use in applications are complementary and equally essential. While some advances are the result of responses to current societal issues, other advances are the result of Blue Sky research, the objective of which is to advance the state of knowledge, independently of the pressing issues of the day, and which will nevertheless be essential for responding to tomorrow's challenges.

Motivated by the potential of spintronics, and because thinking outside the box requires a knowledge of what is already in the box, we are now ready to dive enthusiastically into the following chapters, which focus on the fundamental aspects of the discipline with the aim of providing the basic knowledge to understand most of the underlying physics.

Chapter 2

First Notions About Electron and Spin Transport – CIP-GMR, IEC, AMR

When a system is brought out of equilibrium by an external stimulus, or force, dissipative transport processes cause it to relax back toward equilibrium. If the system is sustained out-of-equilibrium by a stationary force, it reaches a steady out-of-equilibrium state with steady transport processes. If not, the system relaxes back into equilibrium. Here, the word system combines electrons and their angular momenta. External forces can, for example, be caused by (electronic) potential (§2.2), chemical (§3.2), and thermal gradients. Electron and spin angular momentum relaxations are intertwined. These relaxations are due to the alteration of the crystal periodicity, resulting in scattering by crystal defects, lattice vibrations (electron–phonon interaction), electrons (electron–electron interaction, §2.4, §3.2 and §4.1), and magnetic vibrations (electron–magnon). We recall that all the relaxation processes listed above can be spin-dependent because electrons, magnons and phonons can host angular momentum. In this book, we will mostly deal with (spin-dependent) electron–electron interactions.

How electron and spin dynamics are described in spintronics and which type of relaxation mechanism prevails depends on the material's properties and device geometry, as will be presented throughout the text. We will first introduce the types of models that are used, their accuracy and their complexity.

2.1 Types of Model

The models used to describe electron and spin transport are classified into four types, based on how close they are to a quantum mechanical description (figure 3):

- **Phenomenological models**

  This type of model describes empirically the relationship between causes and effects. We will, for example, use such models to illustrate anisotropic magnetoresistance (AMR, §2.6) and giant magnetoresistance (GMR, §2.5), as a forenurmer to the more sophisticated models presented below.

DOI: 10.1051/978-2-7598-2917-0.c002

FIG. 3 – Illustration of several types of models used in spintronics, based on how close they are to a quantum mechanical description.

- **Classical models**
  This type of model tracks deterministically the position of electrons at any time, like in the free electron model, where a position ($\boldsymbol{r}$)- and time (t)- dependent function can be used, for example, to describe the density of electrons $n(\boldsymbol{r}, t)$. An example of a classical model for electron dynamics is the Drude model. It will be briefly evoked in §2.2 and §3.2 because the semiclassical descriptions of spin-independent and spin-dependent transport are connected to it.

- **Semiclassical models**
  In semiclassical models, part of the system is described quantum mechanically. For example, the semiclassical Boltzmann model uses a space-, time-, and momentum($\boldsymbol{k}$)-dependent distribution function $g(\boldsymbol{r}, \boldsymbol{k}, t)$, which corresponds to the Fermi–Dirac distribution function at local equilibrium $f(\boldsymbol{r}, \boldsymbol{k}, t)$, therefore involving the dispersion of Bloch electrons $\varepsilon(\boldsymbol{k})$, where ε is the electron energy. The various physical quantities, like the distribution of conduction electrons and the current, are obtained by performing integrals over the distribution function. The semiclassical Boltzmann transport frame is used to model electron transport like, for example, in the Valet and Fert model for GMR. We will deal with such models in §2.2 and §3.3. Another example of the use of Boltzmann transport is found in §5.2 when deriving the transverse conductivity inherently related to crystal symmetry.

- **Quantum models**
  A quantum model is self-consistent. The state of a system is described by a space- and time-dependent vector, *via* a wavefunction $\psi(\boldsymbol{r}, t)$ when electron transport is at play and *via* a spinor when spin angular momentum is also considered. Its time evolution obeys the Schrödinger equation, which yields the eigenstates $\psi(\boldsymbol{r}, t)$ and the energy eigenvalues $\varepsilon(\boldsymbol{k})$. Physical quantities, like charge and spin current, are represented by observables, and their values are associated with the eigenvalues corresponding to the value of the observable in a given quantum state or eigenstate. A toy quantum model for spin transfer torque (STT) will be presented in §4.2, in which, expressions for the above-mentioned observables are detailed. Quantum mechanics is also used when dealing with the notion of Berry curvature in §5.1, and AMR in §2.6. Other examples of quantum models used in spintronics are the Kubo formalism, which deals with the linear response of an

observable due to a time-dependent perturbation, and the Keldysh formalism for the description of the evolution of a system from a non-equilibrium state.

Next, we will introduce why electron transport can be spin-dependent, using a semiclassical description.

2.2 Two-Current Model, Ohm's Law, Spin-Dependent Mean Free Path

Electrons are split into two categories according to the orientation of their moment/spin. Those two categories are called majority moment(↑)–spin(⇓) electrons and minority moment(↓)–spin(⇑) electrons. It is common to consider the term majority for electrons carrying a magnetic moment $\boldsymbol{m}^{\uparrow}$ 'pointing up', along the magnetization in a ferromagnet (as detailed in §2.3), and thus a spin angular momentum $\boldsymbol{s}^{\Downarrow}$ 'pointing down' from $\boldsymbol{s} = -\boldsymbol{m}/|\gamma|$. Consequently, the minority moment(↓)–spin(⇑) electrons carry a magnetic moment $\boldsymbol{m}^{\downarrow}$, and a spin angular momentum $\boldsymbol{s}^{\Uparrow}$.

We recall that, in the literature, the terms moment $\boldsymbol{m}$ and spin $\boldsymbol{s}$ are sometimes used interchangeably, and the ↑ and ↓ symbols sometimes refer to $\boldsymbol{m}$ or $\boldsymbol{s}$, which can lead to confusion as $\boldsymbol{m}$ and $\boldsymbol{s}$ are of opposite sign. In this book, the term moment refers exclusively to $\boldsymbol{m}$ and spin to $\boldsymbol{s}$; the ↑ and ↓ symbols refer exclusively to the orientation of $\boldsymbol{m}$, and the ⇑ and ⇓ symbols refer exclusively to the orientation of $\boldsymbol{s}$. Note that, in the literature, the + and − symbols are sometimes used for the orientation of $\boldsymbol{s}$.

For now, we also consider that majority moment(↑)–spin(⇓) electrons and minority moment(↓)–spin(⇑) electrons flow in separate channels, according to the two-current model introduced in 1936 by N. F. Mott (figure 4, left). Their (charge) current densities are $\boldsymbol{j}^{\uparrow}$ and $\boldsymbol{j}^{\downarrow}$, respectively.

We note that how the two spin channels mix is described in some of the next sections. Mixing can, for example, occur because of spin-orbit interactions (§2.6), spin-flip scattering (§3.3), and transfer of angular momentum (§4.1).

Because electrons carry a charge and a spin (angular momentum), these two entities naturally flow in matter during electron transport. Therefore, both a charge ($\boldsymbol{J}_e$) and a spin ($\boldsymbol{J}_s$) current density must be considered

$$\boldsymbol{J}_e = \boldsymbol{j}^{\uparrow} + \boldsymbol{j}^{\downarrow} \tag{3}$$

$$\boldsymbol{J}_s = -\left(\boldsymbol{j}^{\uparrow} - \boldsymbol{j}^{\downarrow}\right) \tag{4}$$

In the literature, the spin current density $\boldsymbol{J}_s = -\left(\boldsymbol{j}^{\uparrow} - \boldsymbol{j}^{\downarrow}\right)$ is expressed either in A.m^{-2}, units of charge current density, as here and throughout this textbook, or in J.m^{-2}, units of density of spin angular momentum $\frac{\hbar}{2}$, in which case $\boldsymbol{J}_s = -\frac{\hbar}{2e}\left(\boldsymbol{j}^{\uparrow} - \boldsymbol{j}^{\downarrow}\right)$, resulting in a factor $\frac{\hbar}{2e}$ difference is some of the equations depending on the articles and books consulted. In this textbook, we use $\boldsymbol{Q} = \frac{\hbar}{2e}\boldsymbol{J}_s$.

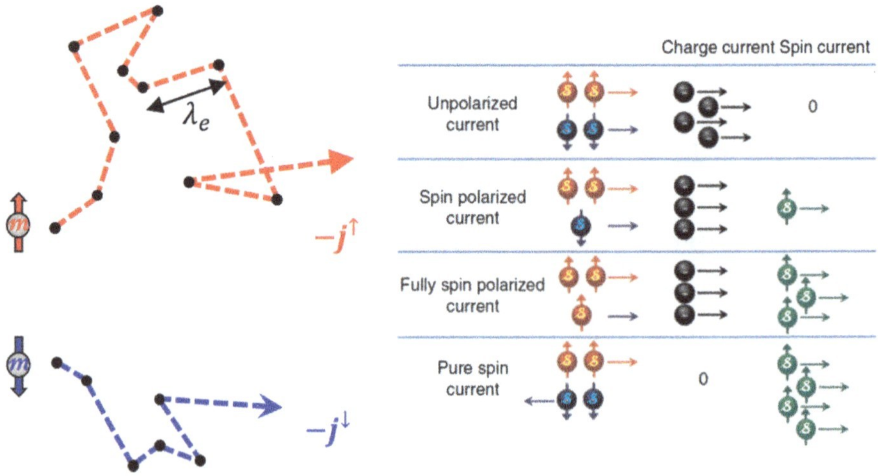

FIG. 4 – (Left) Representation of electron scattering in the two-current model. λ_e is the electron mean free path. Note: to make the representation easy to read, λ_e is shown between two particular collisions. In practice, λ_e is averaged over all collisions. (Right) Schematics of the types of charge and spin currents encountered in spintronics, reprinted with permission from Wiley: Y. F. Feng *et al.*, *WIREs Comput. Mol. Sci.* **7**, e1313 (2017). Copyright 2017.

In the literature, the spin current is also defined either as $\boldsymbol{J}_s = \boldsymbol{j}^{\uparrow} - \boldsymbol{j}^{\downarrow}$, or as $\boldsymbol{J}_s = -(\boldsymbol{j}^{\uparrow} - \boldsymbol{j}^{\downarrow})$, resulting in a factor -1 difference in some of the equations depending on the articles and books consulted. Throughout this textbook, the term spin will refer to $\boldsymbol{s}$ and the spin $\boldsymbol{s}$ current is thus defined as $\boldsymbol{J}_s = \boldsymbol{j}^{\Uparrow} - \boldsymbol{j}^{\Downarrow} = -(\boldsymbol{j}^{\uparrow} - \boldsymbol{j}^{\downarrow})$, because the ↑ and ↓ symbols refer to the orientation of the moment $\boldsymbol{m}$, whereas the ⇑ and ⇓ symbols refer to the orientation of the spin $\boldsymbol{s}$, which points in the opposite direction to $\boldsymbol{m}$ (figure 1).

Charge and spin currents can take very different values, particularly because the net motion of electrons in a given direction is the result of an averaging of instantaneous motions, which may have components in all directions in space. We distinguish the following types of currents (figure 4, right):

- **An unpolarized current**, when the current is made of an equal number of majority moment(↑)–spin(⇓) electrons and minority moment(↓)–spin(⇑) electrons, all flowing in the same direction. In this case, $J_e \neq 0$ and $J_s = 0$.
- **A spin polarized current**, when the current is made of an unequal number of majority moment(↑)–spin(⇓) electrons and minority moment(↓)–spin(⇑) electrons, all flowing in the same direction. In this case, $J_e \neq 0$ and $J_s \neq 0$.
- **A fully spin polarized current**, when the current is made of majority moment(↑)–spin(⇓) electrons only or minority moment(↓)–spin(⇑) electrons only, all flowing in the same direction. In this case, $J_e \neq 0$ and J_s is 100% polarized.

- **A pure spin current**, when the current is made of an equal number of majority moment(↑)–spin(⇓) electrons and minority moment(↓)–spin(⇑) electrons, but flowing in opposite directions. In this case, $J_e = 0$ and J_s is 100% polarized.

The main difference between a charge and a spin current relates to how they obey conservation laws and subsequently to how charge and spin accumulate, as detailed in §3.2. We note that most of the text is based on a one-dimensional description. This simplification was chosen to allow the reader to focus on the actual physical phenomenon at play. Three-dimensional descriptions and the associated complications are discussed in §3.5.

Next, we recall how transport is related to an external stimulus, such as an electric field $\boldsymbol{E}$. It is possible to demonstrate that the semiclassical treatment of electron transport leads to a macroscopic description or Ohm's law [equation (5)], in which the current densities are related to the electric field, and thus to the electric potential, *via* the macroscopic conductivity σ,

$$\boldsymbol{j} = \sigma \boldsymbol{E} \tag{5}$$

with, from the Drude–Sommerfeld model,

$$\sigma = \frac{N(\varepsilon_F)e^2\tau_e}{m_e} = N(\varepsilon_F)e^2 D \tag{6}$$

where $N(\varepsilon_F)$ is the density of states (DOS) at the Fermi level, τ_e is the electronic relaxation time (related to the electron mean free path λ_e *via* the Fermi velocity v_F: $\lambda_e = v_F\tau_e$), and D is the electronic diffusion coefficient.

Using $\boldsymbol{E} = -\boldsymbol{\nabla}\phi$, with ϕ the electric potential, and defining the electrostatic potential as $\mu_e = -e\phi$, the relation between current density and electrostatic potential can be obtained from equation (5)

$$\boldsymbol{j} = \sigma \frac{\boldsymbol{\nabla}\mu}{e} \tag{7}$$

here, with $\mu = \mu_e$. This type of macroscopic description remains valid when spin-dependent coefficients are involved [equation (8) and §2.3] and when electrostatic potentials are extended to electrochemical potentials [equation (22)], when $\mu = \mu_e + \mu_n$. The validity of this assertion in the most general case will be discussed in more detail in §3.2. The correct term is electrostatic potential energy because it is a potential energy, the unit of which is joule, but it is customary to use the abbreviated term electrostatic potential, just as the term electrochemical potential describes a potential energy.

In the next part, we will explain why τ_e (or λ_e) is spin-dependent in ferromagnets, therefore making $\boldsymbol{j}^\uparrow \neq \boldsymbol{j}^\downarrow$ [from equations (6) and (7)], resulting in a spin-dependent generalized Ohm's law

$$\boldsymbol{j}^{\uparrow(\downarrow)} = \sigma^{\uparrow(\downarrow)} \frac{\boldsymbol{\nabla}\mu^{\uparrow(\downarrow)}}{e} \tag{8}$$

Notation: equation (8) is two concatenated equations: $\boldsymbol{j}^\uparrow = \sigma^\uparrow \frac{\boldsymbol{\nabla}\mu^\uparrow}{e}$ and $\boldsymbol{j}^\downarrow = \sigma^\downarrow \frac{\boldsymbol{\nabla}\mu^\downarrow}{e}$. This type of notation is used throughout.

2.3 Band Structures, Spin-Dependent Fermi Surface & Density of State

In metals, electronic conduction occurs at or near the Fermi surface, *i.e.*, for $\varepsilon = \varepsilon_F$ (figures 5 and 6). Due to band splitting, ferromagnetic metals have different Fermi surfaces for majority moment(↑)–spin(⇓) electrons and minority moment(↓)–spin (⇑) electrons (figure 6), making electron transport spin-dependent. We recall that band splitting causes a gain of Coulomb energy, because some electrons are unpaired, at the expense of a loss of kinetic energy, because electrons move up in energy. The overall energy balance is favorable for densely packed electrons when a small loss of kinetic energy leads to a large gain of Coulomb energy. In turn, the delocalized *s*-electrons have no tendency for band magnetism, in contrast to the *5f*-electrons of the early actinides and the *3d*-electrons of the late transition metals, which possess sufficiently narrow bands. Narrower bands favor localized (sometimes almost atomic) magnetism, like the late actinides and rare earth compounds. In this book, we focus on band magnetism [see also Stoner criterion in J. M. D. Coey, *Magnetism and Magnetic Materials*, Cambridge University Press (2010)].

The relation between exchange splitting and spin-dependent transport is presented next, based on the partial densities of states DOS, $N^{\uparrow(\downarrow)}$ (figure 7). A DOS represents the number of continuum states in an infinitesimally small energy interval $\varepsilon + d\varepsilon$. In *3d*-transition metals, spins contributing to transport are split into two types:

- **itinerant (delocalized) spins** carried by light *4s*-electrons, corresponding to the vertically elongated parts of the DOS in figure 7, which are not spin-split in energy, and
- **localized spins** carried by heavy *3d*-electrons, corresponding to the horizontally elongated parts of the DOS in figure 7, which are spin-split in energy for the Fe, Ni, and Co ferromagnets in the figure and unsplit for non-magnetic Cu, where the *3d*-bands are full.

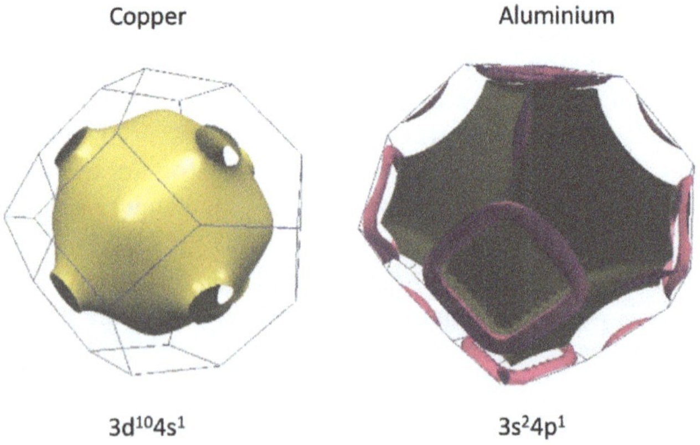

FIG. 5 – The Fermi surfaces of two non-magnetic metals: copper and aluminium. From http://www.phys.ufl.edu/fermisurface/.

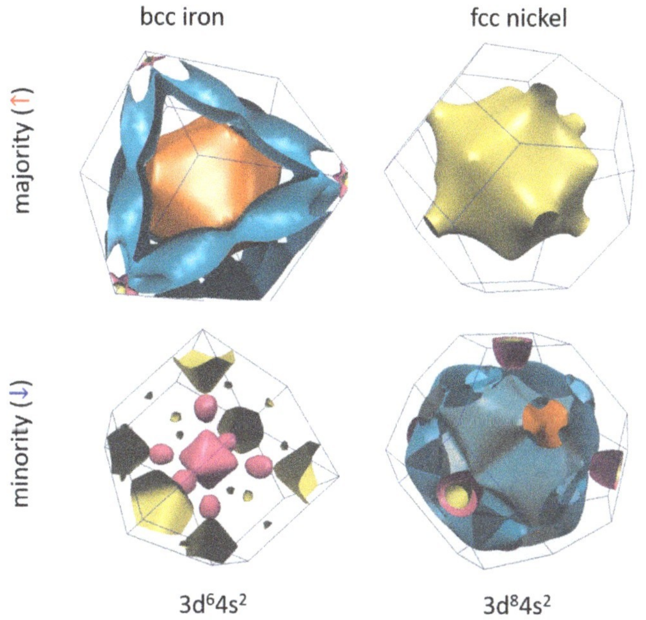

FIG. 6 – The Fermi surfaces of (Left) bcc-iron (body-centered cubic – Fe) and (Right) fcc-nickel (face-centered cubic – Ni) for majority moment(↑)–spin(⇓) electrons and minority moment(↓)–spin(⇑) electrons. From http://www.phys.ufl.edu/fermisurface/. The colors correspond to the contributions of several bands.

As a result, for spin-split ferromagnets, the DOS at the Fermi level is spin-dependent, for example, $N^{\uparrow}(\varepsilon_F) < N^{\downarrow}(\varepsilon_F)$ for hcp-Co.

The terms 'light' and 'heavy' come from the effective mass m_e^* theory used to simplify band structures by modelling the behaviour of a free electron with this mass: $\varepsilon(k) = \frac{\hbar^2 k^2}{2m_e^*}$ and $m_e^* = \hbar^2 / \frac{\partial^2 \varepsilon}{\partial k^2}$.

Note that, the spin polarization obtained from the DOS at the Fermi level $\frac{N^{\downarrow}(\varepsilon_F) - N^{\uparrow}(\varepsilon_F)}{N^{\downarrow}(\varepsilon_F) + N^{\uparrow}(\varepsilon_F)}$ depends on how the bands are split, explaining why hcp-Co and bcc-Fe have opposite DOS spin polarization (figure 7). Note also that, transport ($\boldsymbol{j}$) relates to processes occurring near the Fermi level whereas magnetism relates to properties up to the Fermi level, *i.e.*, magnetization (M) relates to the integral from the bottom of the band to the Fermi level $M = \mu_{\mathrm{B}} \int_{-\infty}^{\varepsilon_F} f(\varepsilon) N^{\uparrow}(\varepsilon)\, d\varepsilon - \int_{-\infty}^{\varepsilon_F} f(\varepsilon) N^{\downarrow}(\varepsilon)\, d\varepsilon$, with $f(\varepsilon) = \frac{1}{\exp\left(\frac{\varepsilon - \mu_n}{k_B T}\right) + 1}$ the Fermi–Dirac distribution and μ_n the chemical potential defined in §3.2. The majority moment(↑)–spin(⇓) electrons can be in the minority near the Fermi level compared to the minority moment(↓)–spin(⇑) electrons which are a majority, see *e.g.*, hcp-Co in figure 7.

How the vertically elongated unsplit partial DOS of the itinerant *s*-electrons and the horizontally elongated spin-split partial DOS of the localized *d*-electrons are represented in a simplified manner are shown in figure 8.

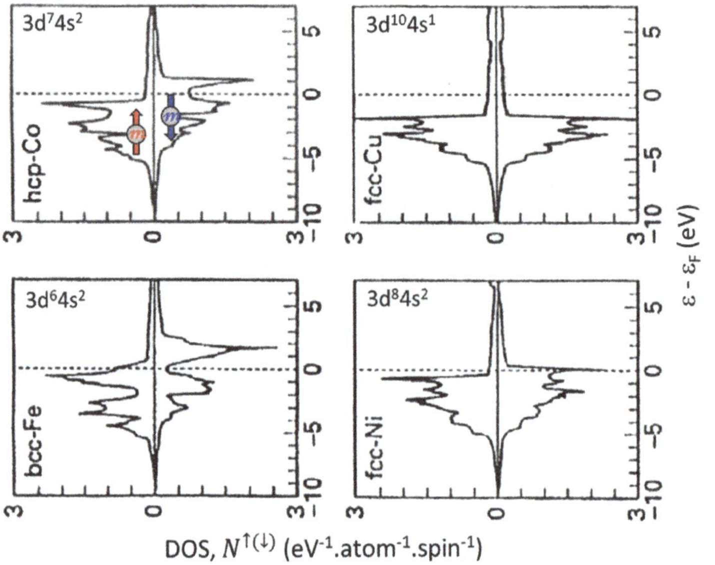

FIG. 7 – Energy($\varepsilon - \varepsilon_F$)-dependence of the partial density of state DOS, $N^{\uparrow(\downarrow)}$ of *3d* transition metals for majority moment(↑)–spin(⇓) electrons (left quadrant in all 4 graphs) and minority moment(↓)–spin(⇑) electrons (right quadrant), obtained by *ab-initio* calculations. The Fermi level is represented by the horizontal dashed line at $\varepsilon - \varepsilon_F = 0$. Note that, here, N is given in $eV^{-1}.atom^{-1}.spin^{-1}$ compared to $J^{-1}.m^{-3}$ in the text. Adapted with permission from Springer Nature: J. Mathon & A. Umerski, Theory of Tunneling Magnetoresistance in J. L. Moran-Lopez (eds) *Physics of Low Dimensional Systems*, Springer, Boston. Copyright 2001.

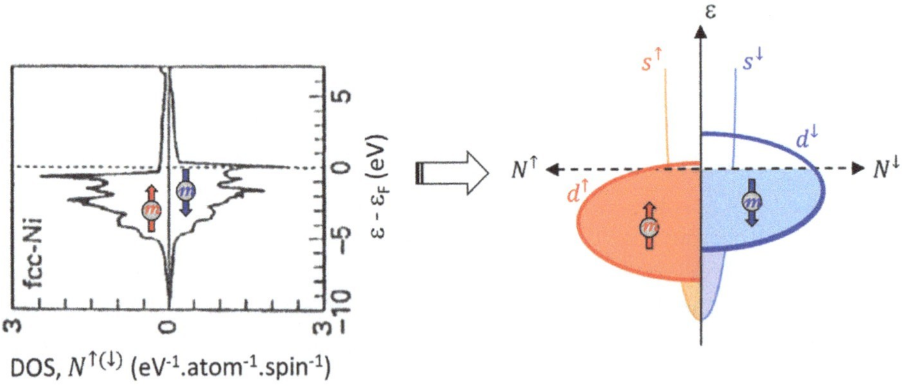

FIG. 8 – Illustration of how the energy($\varepsilon - \varepsilon_F$)-dependence of the partial density of state DOS, $N^{\uparrow(\downarrow)}$ is represented in a simplified manner. The colors illustrate that the *s*- and *d*-bands are filled up to the Fermi level indicated by the dashed line. The image on the left is adapted with permission from Springer Nature: J. Mathon & A. Umerski, Theory of Tunneling Magnetoresistance in J. L. Moran-Lopez (eds) *Physics of Low Dimensional Systems*, Springer, Boston. Copyright 2001.

2.4 Spin-Dependent Scattering, the *sd* Model

We now introduce the *sd* model for which: (i) the current is carried by the *s*-electrons, and (ii) elastic scattering between the itinerant *s*-electrons and the localized *d*-electrons dominates.

We consider electron–electron scattering only, and the two-current model stipulates that majority moment(↑)–spin(⇓) electrons and minority moment(↓)–spin(⇑) electrons flow in separate channels (§2.2). It follows that a $s^{\uparrow}$-electron can be scattered by a $s^{\uparrow}$-electron or a $d^{\uparrow}$-electron, and that a $s^{\downarrow}$-electron can be scattered by a $s^{\downarrow}$-electron or a $d^{\downarrow}$-electron.

According to Fermi's golden rule, the scattering probability $P^{i\to f}$ of an electron in an initial state $|i\rangle$ to a final state $|f\rangle$ is determined by the DOS $N^f(\varepsilon_F)$ allowed in the final state, after scattering $P^{i\to f} \propto \langle f|\mathcal{H}|i\rangle^2 N^f(\varepsilon_F)$, where $\mathcal{H}$ is the scattering Hamiltonian. Based on the largest DOS at the Fermi level $N^{\downarrow}(\varepsilon_F)$ for the minority moment(↓)–spin(⇑) electrons $d^{\downarrow}$, it is possible to conclude that the most efficient channel for scattering corresponds to an itinerant $s^{\downarrow}$-electron scattered by a localized $d^{\downarrow}$-electron (figure 9), therefore making the electronic scattering rate spin-dependent: $\tau_e^{\uparrow} > \tau_e^{\downarrow}$. As an example, for Co, we have $\tau_e^{\uparrow} \sim 10\tau_e^{\downarrow}$, with $\lambda_e^{\uparrow} \sim 10$ nm and $\lambda_e^{\downarrow} \sim 1$ nm (Note: $\lambda_e = v_F \tau_e$, with $v_F \sim 2 \times 10^6\,\mathrm{m.s^{-1}}$).

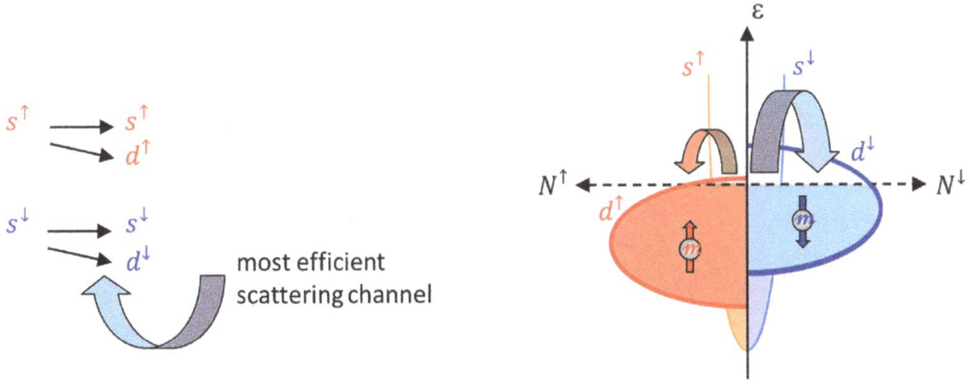

FIG. 9 – Representation of electron scattering in the two-current model. Based on the DOS at the Fermi level: (i) scattering between the itinerant *s*-electrons and the localized *d*-electrons is more efficient than scattering between *s*-electrons, no matter the conduction channel considered (↑ or ↓), and (ii) this type of *sd* scattering is most efficient for minority moment(↓)–spin(⇑) electrons, because there are more $d^{\downarrow}$-states to receive them.

Because τ_e (or λ_e) is spin-dependent, it naturally follows from equations (6) and (7), that $\boldsymbol{j}^{\uparrow} \neq \boldsymbol{j}^{\downarrow}$, resulting in the spin-dependent generalized Ohm's law anticipated in equation (8) and repeated here: $\boldsymbol{j}^{\uparrow(\downarrow)} = \sigma^{\uparrow(\downarrow)} \frac{\nabla \mu^{\uparrow(\downarrow)}}{e}$ (two equations).

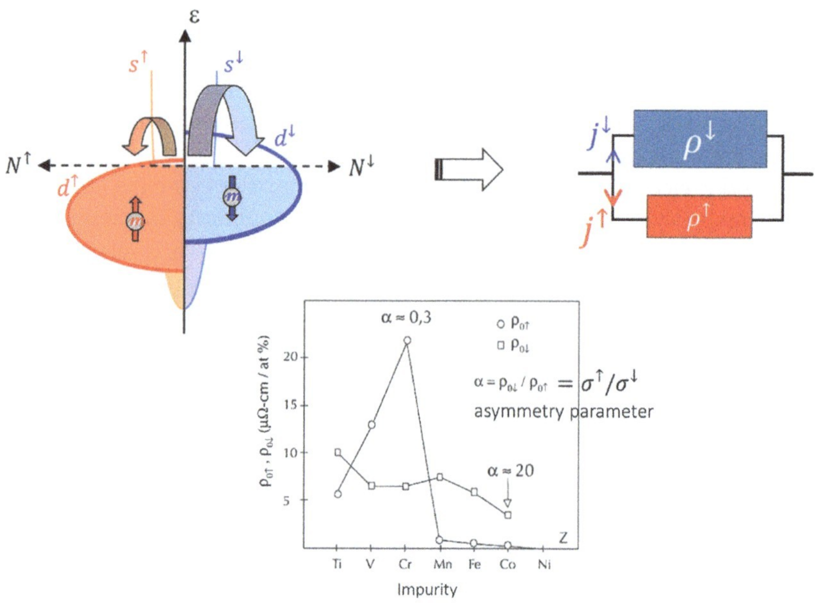

FIG. 10 – (Top) Illustration of how asymmetric electron scattering, due to spin-dependent DOS, is represented in an equivalent circuit of two parallel resistors. (Bottom) Asymmetry parameter $\alpha = \sigma^{\uparrow}/\sigma^{\downarrow}$ experimentally determined for impurities of *3d* transition metals in Ni. Adapted with permission from EDP Sciences: A. Fert, *Reflets Phys.* **15**, 5 (2009).

Spin-dependent electron scattering due to band splitting is also referred to as asymmetric electron scattering, and the resulting spin-dependent generalized Ohm's law [equation (8)] can be represented in an equivalent circuit of two parallel resistors or resistivities, wherein the incoming charge current $\boldsymbol{J}_e = \boldsymbol{j}^{\uparrow} + \boldsymbol{j}^{\downarrow}$ splits in two (figure 10). A high resistivity value $\rho^{\downarrow} = 1/\sigma^{\downarrow}$ accounts for a large number of *sd* scattering events in the minority moment($\downarrow$)–spin($\Uparrow$) electron channel hosting the $\boldsymbol{j}^{\downarrow}$ current, and a smaller resistivity value $\rho^{\uparrow} = 1/\sigma^{\uparrow}$ is used for less *sd* scattering in the majority moment ($\uparrow$)–spin($\Downarrow$) electrons channel hosting the $\boldsymbol{j}^{\uparrow}$ current. The total resistance of this equivalent circuit is expressed as

$$\rho = \frac{\rho^{\uparrow}\rho^{\downarrow}}{\rho^{\uparrow} + \rho^{\downarrow}} \tag{9}$$

It can be rearranged to account for the spin polarization β of the current at the Fermi level

$$\rho = \rho^{*}(1 - \beta^{2}) \tag{10}$$

with $\rho^{\uparrow(\downarrow)} = 2\rho^{*}(1 - (+)\beta)$, and

$$\beta = \frac{\rho^{\downarrow} - \rho^{\uparrow}}{\rho^{\downarrow} + \rho^{\uparrow}} = \frac{\alpha - 1}{\alpha + 1} \tag{11}$$

where α is the asymmetric scattering parameter

$$\alpha = \frac{\rho^{\downarrow}}{\rho^{\uparrow}} \tag{12}$$

We will see other examples of the use of β later, in the context of CIP-GMR (§2.5), and of the injection of a current across an interface (§3.2) and the subsequent transfer of angular momentum (§4.3). The spin polarization of the current relates to the parameter α also known as the spin asymmetry parameter as it relates to asymmetric electron scattering due to the spin-dependent DOS at the Fermi level. It is a material/element-dependent parameter, which can be determined experimentally (figure 10).

Note that, the generic principle of the tunnel magnetoresistance effect (TMR) is based on spin-dependent DOS at the Fermi level, as it drives the tunneling probability across the tunnel junction of a ferromagnet/tunnel barrier/ferromagnet trilayer. A toy quantum model accounting for wavefunction propagation through a potential barrier depicts the phenomenon well. More subtle orbital-dependent tunneling is at play in the latest TMR devices. The principles of TMR are extensively described in the literature [S. Yuasa *et al.*, *J. Phys. D: Appl. Phys.* **40**, R337 (2007); W. H. Butler, *Sci. Technol. Adv. Mater.* **9**, 014106 (2008)]. TMR devices are now extensively used in memory and sensor applications (§2.5), and still require developments at the engineering level but not so much at the basic level, except for the newest materials (chapter 7).

2.5 From Impurity Scattering to Heterostructures – CIP-GMR, IEC

In the previous section, we discussed how asymmetric electron scattering influences transport in a mono-element ferromagnet. In this section, we will consider how asymmetric electron scattering influences transport, first, in ternary ferromagnetic alloys and then, in heterostructures made of a ferromagnet/non-magnet/ferromagnet trilayer [A. Fert, *Reflets Phys.* **15**, 5 (2009)], and discuss a related effect known as the interlayer exchange coupling.

■ Element doping control over a stack's resistance – the example of ternary alloys

We consider a ferromagnet made of Ni, doped with Co and Rh elements having spin asymmetry parameters larger and smaller than 1, respectively: $\alpha_{\mathrm{Co}} = \rho^{\downarrow}_{\mathrm{Co}}/\rho^{\uparrow}_{\mathrm{Co}} > 1$ and $\alpha_{\mathrm{Rh}} = \rho^{\downarrow}_{\mathrm{Rh}}/\rho^{\uparrow}_{\mathrm{Rh}} < 1$ (figure 11, Type 1, Ni(Co-Rh) alloy). Let us compare it to the case of the same Ni ferromagnet but doped with Au and Co elements, both having spin asymmetry parameters larger than 1: $\alpha_{\mathrm{Co}} = \rho^{\downarrow}_{\mathrm{Co}}/\rho^{\uparrow}_{\mathrm{Co}} > 1$ and $\alpha_{\mathrm{Au}} = \rho^{\downarrow}_{\mathrm{Au}}/\rho^{\uparrow}_{\mathrm{Au}} > 1$ (figure 11, Type 2, Ni(Au-Co) alloy).

In the case of the Ni(Co_{1-x}-Rh_x) alloy, the electrons' flow is altered in the two spin channels. In the majority moment(↑)–spin(⇓) electrons channel, electrons are (additionally)-scattered by the Rh atoms (on top of the initial scattering by Ni

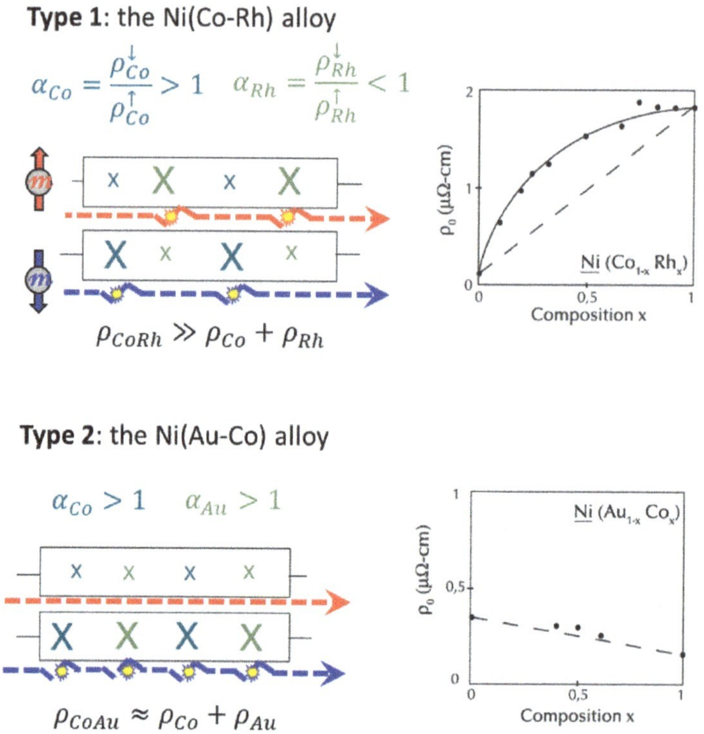

FIG. 11 – (Left) Representation of spin-dependent scattering in ternary alloys, made of a Ni host and two impurities. Type 1: the two impurities (Co and Rh) have asymmetric scattering parameters larger and smaller than 1, respectively. Consequently, they alter electron transport in different spin channels, making both channels more resistive. Type 2: the two impurities (Co and Au) both have asymmetric scattering parameters larger than 1, consequently, they alter electron transport in the same spin channel, making electron flow efficient in the unaltered channel – here, the channel for majority moment(↑)–spin(⇓) electrons. (Right) Experimentally determined data. Adapted with permission from EDP Sciences: A. Fert, *Reflets Phys.* **15**, 5 (2009).

atoms), as $\alpha_{\text{Rh}} = \rho^{\downarrow}_{\text{Rh}}/\rho^{\uparrow}_{\text{Rh}} < 1$. In the minority moment(↓)–spin(⇑) electrons channel, electrons are (additionally)-scattered by the Co atoms, as $\alpha_{\text{Co}} = \rho^{\downarrow}_{\text{Co}}/\rho^{\uparrow}_{\text{Co}} > 1$ (figure 11, Type 1). As a result, the overall resistivity of the ternary alloy is increased compared to expectations, if spin-dependent scattering were discarded.

In contrast, in the case of the Ni(Au_{1-x}-Co_x) alloy, the electrons' flow is altered in one spin channel only. In the majority moment(↑)–spin(⇓) electrons channel, electrons are not (additionally)-scattered, and in the minority moment(↓)–spin(⇑) electrons channel, electrons are (additionally)-scattered by both the Co and the Au atoms, because $\alpha_{\text{Co}} = \rho^{\downarrow}_{\text{Co}}/\rho^{\uparrow}_{\text{Co}} > 1$ and $\alpha_{\text{Au}} = \rho^{\downarrow}_{\text{Au}}/\rho^{\uparrow}_{\text{Au}} > 1$ (figure 11, Type 2). As a result, the overall resistivity of the ternary alloy is almost unchanged.

Note that, the reshaping of the DOS of the Ni ferromagnet due to the presence of dopants explains the use of asymmetric parameters α_{Rh} and α_{Au} associated with non-magnetic dopants like Rh and Au. In fact, the presence of a non-magnetic impurity in a magnetic metal host will create a virtual bond state, with a DOS of finite width in energy. This state will hybridize with its magnetic environment. Depending on the position of the virtual bond state with respect to the Fermi level of the magnetic host, electrons are removed or added at the Fermi level, enhancing or diminishing the magnetic asymmetry of the ferromagnet, and conversely conferring the impurity a magnetic character, with momentum in the opposite or same direction to the host, respectively [J. Friedel, *Nuovo Cim. Suppl.* **2**, 287-3 (1958); R. Zeller, *J. Phys. F: Met. Phys.* **17**, 2123 (1987)].

■ Relative direction of magnetization control over a stack's resistance – CIP-GMR

In the section above, we have seen that the spin-dependent resistance of a stack can be controlled by the nature of dopants. The initial idea underlying the giant magnetoresistance effect (GMR) is to 'replace' the dopants with ferromagnetic layers (Fs) and the ternary alloy by a F/non-magnet(N)/F trilayer. The magnetizations of the two Fs can be manipulated independently in such a way that they can be set pointing in opposite directions, resulting in the antiparallel (AP) state, or in the same direction, resulting in the parallel (P) state (figure 12). The current is in-plane (CIP). Note that, the magnetization of a ferromagnet can be manipulated by an external magnetic field or an electric current through spin torques (chapter 4). In addition, the two layers can be manipulated independently, if for example, their coercive (reversal) field differ or if one of the two layers is pinned by exchange bias in spin-valve devices (§2.5) [J. Nogues *et al.*, *J. Magn. Magn. Mater.* **192**, 203 (1999)].

In the $F^{\uparrow}/N/F^{\downarrow}$ AP state, the electron flow is altered in the two spin channels due to bulk scattering upon crossing the $F^{\uparrow}/N$ and $N/F^{\downarrow}$ interfaces to enter the $F^{\uparrow}$ and $F^{\downarrow}$ layers (figure 12, top). In the majority moment(↑)–spin(⇓) electrons channel, electrons are (additionally)-scattered near the $N/F^{\downarrow}$ interface since $\alpha_{F^{\downarrow}} = \rho^{\uparrow}/\rho^{\downarrow} > 1$, and in the minority moment(↓)–spin(⇑) electrons channel, electrons are (additionally)-scattered near the $F^{\uparrow}/N$ interface as $\alpha_{F^{\uparrow}} = \rho^{\downarrow}/\rho^{\uparrow} > 1$. As a result, the overall resistivity of the AP state is 'large'. This behavior is equivalent to the type-1 ternary alloy discussed in the previous section (figure 11). From the corresponding equivalent circuit (figure 12, top right), the resistivity of the AP state is

$$\rho^{AP} = \frac{\rho^{\uparrow} + \rho^{\downarrow}}{2} \tag{13}$$

We considered that the two Fs are the same here. Different Fs lead to different asymmetric parameters and different spin-dependent resistivities, making the formula more complex, but leaving the physical explanation unchanged.

In the $F^{\uparrow}/N/F^{\uparrow}$ P state, the electrons' flow is altered in one spin channel only due to scattering near the $F^{\uparrow}/N$ and $N/F^{\uparrow}$ interfaces (figure 12, bottom). In the majority moment(↑)–spin(⇓) electrons channel, electrons are not (additionally)-scattered, whereas in the minority moment(↓)–spin(⇑) electrons channel, electrons are

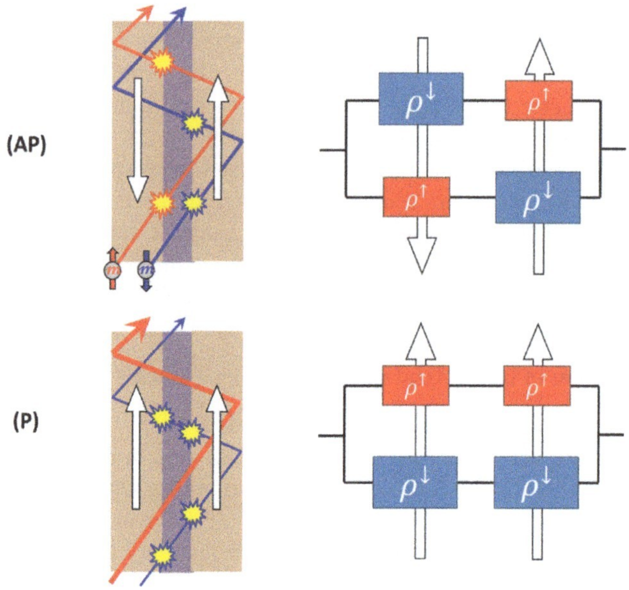

FIG. 12 – (Left) Schematics of the current-in-plane (CIP) giant magnetoresistance (GMR) effect, which is a prototypical example of how asymmetric electron scattering influences electron transport – due to spin-dependent DOS and subsequent spin-dependent scattering near interfaces in a ferromagnet/non-magnet/ferromagnet trilayer. Specular reflection is considered at the farthest interfaces. (Right) Corresponding equivalent circuit. Adapted with permission from Springer Nature: P. Grünberg *et al.*, Metallic Multilayers, in Y. Xu *et al.* (eds) *Handbook of Spintronics*, Springer, Dordrecht (2015). Copyright 2015.

(additionally)-scattered near both interfaces as $\alpha_{F\uparrow} = \rho^{\downarrow}/\rho^{\uparrow} > 1$. As a result, the overall resistivity of the P state is 'smaller'. This behavior is equivalent to the type-2 ternary alloy discussed in the previous section (figure 11). From the corresponding equivalent circuit (figure 12, bottom right), the resistivity of the P state is

$$\rho^P = \frac{2\rho^{\uparrow}\rho^{\downarrow}}{\rho^{\uparrow} + \rho^{\downarrow}} \tag{14}$$

The magnetoresistance, here known as the current-in-plane giant magnetoresistance, CIP-GMR reads

$$\text{CIP-GMR} = \frac{\rho^{AP} - \rho^P}{\rho^{AP}} = \left(\frac{\rho^{\uparrow} - \rho^{\downarrow}}{\rho^{\uparrow} + \rho^{\downarrow}}\right)^2 = \left(\frac{\alpha - 1}{\alpha + 1}\right)^2 \tag{15}$$

CIP-GMR depends on spin-dependent asymmetric scattering α due to spin-dependent DOS. The important characteristic length for CIP-GMR is the spin-dependent electron mean free path, $\lambda_e^{\uparrow(\downarrow)}$ (§2.3). Neither spin-flip scattering (chapter 3) nor spin mixing (§2.6) is needed to create CIP-GMR. Those effects can alter CIP-GMR, but they are not necessary to trigger the effect. Conversely, we will see in chapter 3 that current perpendicular-to-plane (CPP)-GMR is driven by spin-flip scattering and that the important characteristic length for CPP-GMR is thus the spin diffusion length l^*_{sf}. The fact that $l^*_{sf} \gg \lambda_e^{\uparrow(\downarrow)}$ (roughly a factor of 10 larger) marks an essential difference between CPP- and CIP-GMR.

The terminology 'giant' refers to the fact that these types of magnetoresistance effects, demonstrated in the late 1980ies – early 1990ies, are much larger than the anisotropic magnetoresistance effect discovered much earlier, in the late 1850ies – modelled in the early 1970ies and due to spin-orbit spin mixing (§2.6). A. Fert and P. Grünberg were awarded the Nobel prize in physics in 2007 "for the discovery of the giant magnetoresistance effect".

■ Note about the interlayer exchange coupling (IEC)

We next present interlayer exchange coupling (IEC), which is a companion effect to CIP-GMR. It is also closely related to Ruderman–Kittel–Kasuya–Yosida (RKKY) interaction between magnetic impurities.

For CIP-GMR, the thickness d_N of the non-magnetic (N) spacer layer is fixed and the relative alignment of the ferromagnetic layers is varied, for example, by use of an external magnetic field or an electric current *via* spin torque (chapter 4), leading to a change in resistance (figure 12 and corresponding text).

Conversely, for IEC, d_N is varied, leading directly to a change in the relative orientation of the magnetization of the ferromagnetic layers and subsequently to a change in resistance (figure 13). Indeed, in the case of the P state (figure 13, top right), minority moment($\downarrow$)–spin($\Uparrow$) electrons can be trapped between the two ferromagnets due to constructive scattering near the $F^{\uparrow}/N$ and $N/F^{\uparrow}$ interfaces. It results in a one-dimensional quantum well behavior.

For a simplified explanation, consider quantum well states (QWS) of quantized energy levels

$$\varepsilon_n = \frac{\hbar^2 k_\perp^2}{2m_e} = \frac{\hbar^2 n^2 \pi^2}{2m_e d_N^2} \tag{16}$$

where $k_\perp$ is the transverse (perpendicular-to-plane) wavevector and n is an integer. "For increasing d_N, the levels move downward. When a level crosses the Fermi energy ε_F, the QWS is populated, and the total energy increases. When the QWS level moves further below ε_F, the energy again decreases until the next level approaches ε_F" – quoted from P. Grünberg *et al.*, Metallic Multilayers, in Y. Xu *et al.* (eds) *Handbook of Spintronics*, Springer (2015). Consequently, for P alignment, the energy oscillates with d_N.

In the case of the AP state (figure 13, top left), no such quantization effect occurs because, for a given spin-electron channel, spin-dependent scattering at interfaces is not active in a collaborative manner at both interfaces. Therefore, electrons have a 'classical' behavior and the energy of the AP state is independent of d_N.

Because the energy of the P state oscillates around the energy of the AP state, and in order to minimize energy, the alignment switches between P and AP states when d_N increases, meaning that the sign of the coupling oscillates between positive and negative. These oscillations result in subsequent oscillations of the CIP-GMR

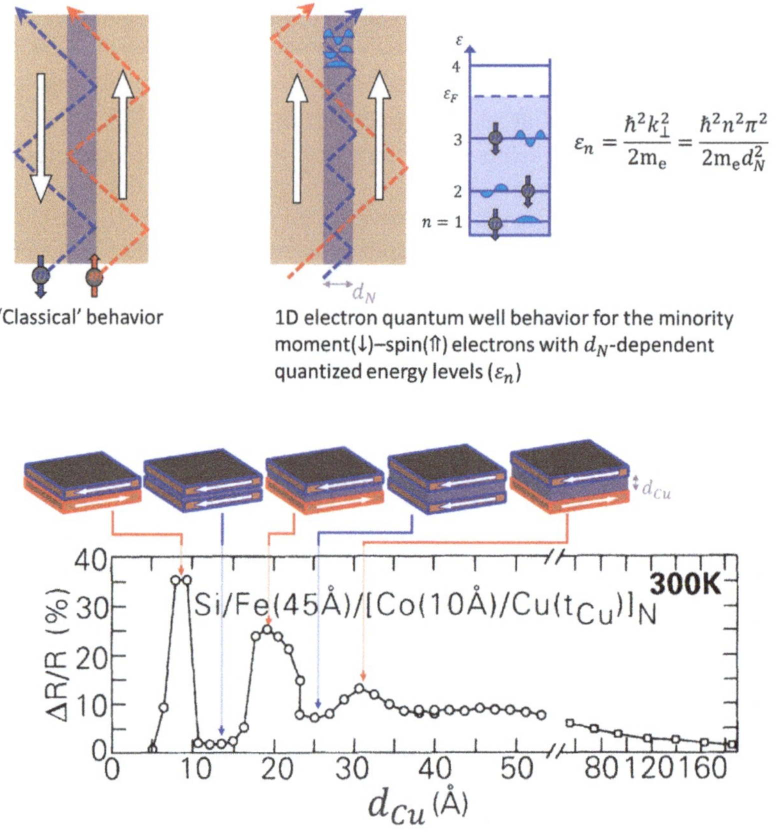

'Classical' behavior

1D electron quantum well behavior for the minority moment(\downarrow)–spin(\Uparrow) electrons with d_N-dependent quantized energy levels (ε_n)

$$\varepsilon_n = \frac{\hbar^2 k_\perp^2}{2m_e} = \frac{\hbar^2 n^2 \pi^2}{2m_e d_N^2}$$

FIG. 13 – (Top) Illustration of the interlayer exchange coupling (IEC). Adapted with permission from Springer Nature: P. Grünberg *et al.*, Metallic Multilayers, in Y. Xu *et al.* (eds) *Handbook of Spintronics*, Springer, Dordrecht (2015). Copyright 2015. (Bottom) Experimental data showing the oscillatory behavior of the GMR as a function of the non-magnetic layer thickness in a ferromagnet/non-magnet/ferromagnet trilayer, due to the oscillatory sign of IEC. Adapted with permission from the American Physical Society: S. S. P. Parkin, *Phys. Rev. Lett.* **66**, 2152 (1991). Copyright 1991.

between a highly resistive AP state and a less-resistive P state (figure 13, bottom). Note that, the present reasoning for IEC does not require a current to be applied in the trilayer. The application of a current will only allow measuring the resistance of the trilayer. It will not have any effect on the relative alignment of the magnetizations of the layers at first order, unless transfer of angular momentum takes place, which will be the object of chapter 4.

To describe real systems beyond the simplified picture presented above, refined models must take into account the reshuffling of the spacer layer DOS due to the change from the initial behaviour of the bulk to that of the quantum well [M. D. Stiles, *Phys. Rev. B* **48**, 7238 (1993); P. Bruno, *Phys. Rev. B* **52**, 411 (1995)]. This change is dependent on d_N, as the properties of the quantum well are determined by d_N. It can be shown that electrons have to be taken out or added in at the Fermi level, alternatively depending on d_N. Therefore, the total energy on filling the states up to the Fermi level oscillates with d_N. The periods of the oscillations were calculated to be $\pi/k_{F,es}$, with $k_{F,es}$ some extremal spanning vectors of the Fermi surface of the spacer layer. We note that, most of the multiple oscillations due to k_F cancel out on integration over the Fermi surface, leaving only the oscillations due to the extrema, $k_{F,es}$.

■ Numbers

The purpose of this part is to provide some orders of magnitude for the effects discussed above, in the context of the applications in which they are at play. Because the present book is focused on basic effects, readers may complement their knowledge on spintronic applications by consulting for example E. du Tremolet de Lacheisserie *et al.* (eds) *Magnetism: II. Materials and Applications*, EDP Sciences (2002).

Initial GMR values were around few percent (figure 14, left), but typical best values for GMR now reach few tens of percent at 300K. This is mostly due to the different mechanisms and related length scales involved, namely, the spin-dependent mean free path $\lambda_e^{\uparrow(\downarrow)}$ for CIP-GMR (§2.4) and the spin-diffusion length l_{sf}^* for

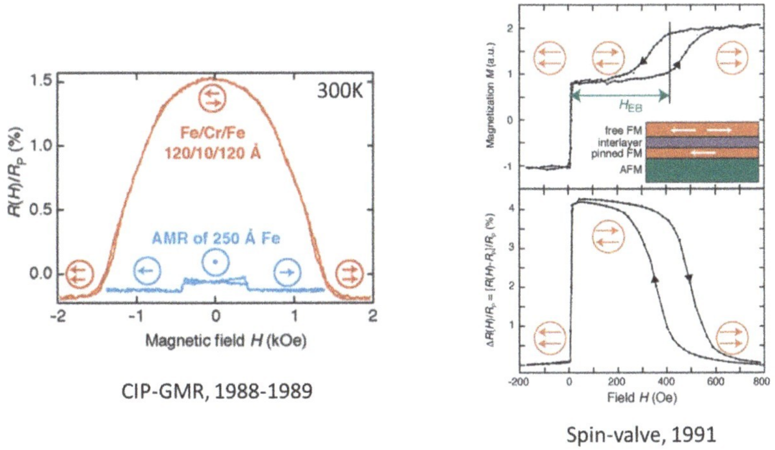

FIG. 14 – (Left) Experimental data of the GMR effect in an Fe/Cr/Fe trilayer with in-plane anisotropy compared to data of the AMR effect for an Fe monolayer of equivalent total thickness. Reprinted with permission from the American Physical Society: G. Binasch *et al.*, *Phys. Rev. B* **39**, 4828 (1989). Copyright 1989. See also M. N. Baibich *et al.*, *Phys. Rev. Lett.* **61**, 2472 (1988) for related results in [Fe/Cr] multilayers with out-of-plane anisotropy. (Right) Experimental data of the magnetic behavior and electric GMR effect in a spin-valve structure, wherein one of the ferromagnetic layers is pinned by exchange bias coupling to an antiferromagnet. Reprinted with permission from Elsevier: B. Dieny, *J. Magn. Magn. Mater.* **136**, 335 (1994). Copyright 1994.

CPP-GMR (§3.2). Larger values of few hundred of percent can be obtained in TMR devices, relying on more selective orbital-dependent tunneling. The other commonly encountered and historically much older anisotropic magnetoresistance effect known as AMR comes from spin-orbit scattering and gives magnitudes of few percent. The impact of spin-orbit interactions and the AMR effect will be discussed in the next section (§2.6).

Note that, the application of those magnetoresistance effects was made possible, thanks to the invention of spin-valves (figure 14, right). One of the ferromagnetic layers is used as a reference, through pinning by exchange bias coupling to an antiferromagnet [J. Nogues *et al.*, *J. Magn. Magn. Mater.* **192**, 203 (1999)], and the other ferromagnetic layer is used for sensing, as it is free to rotate when subject to an external field, *e.g.*, stray fields from encoded bits in a hard disk drive for data storage (figure 15, top). The discovery and deep understanding of those basic effects, along with thorough engineering, ended up, for example, in the fabrication and use of several billions of sensors. Coupled to the spin transfer effects (chapter 4) for writing, TMR is also in play for reading in some of the latest memory (MRAM) applications (figure 15, bottom).

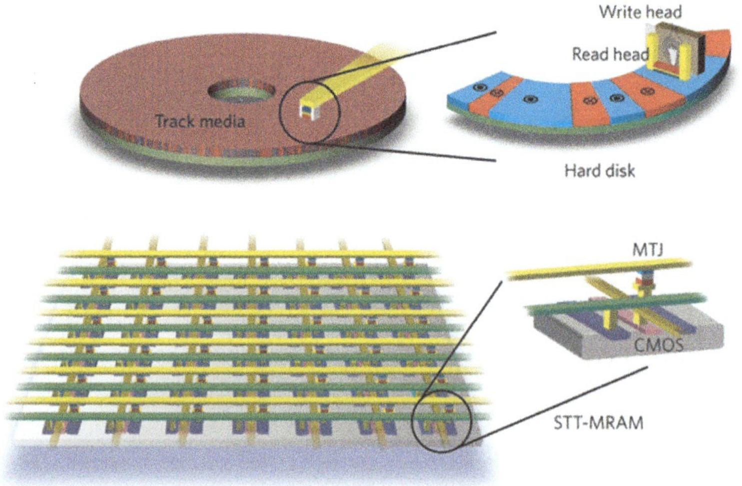

FIG. 15 – (Top) Schematics of a magnetic disk with a read head on top for use in hard disk drives. (Bottom) Schematics of a magnetic memory made of STT-MRAM cells. Reprinted with permission from Springer Nature: S. S. P. Parkin *et al.*, *Nat. Nanotech.* **10**, 195 (2015). Copyright 2015.

2.6 Spin Mixing and Spin-Orbit Interactions – AMR

Up to now, we considered that the majority moment(↑)–spin(⇓) electrons and minority moment(↓)–spin(⇑) electrons flow in channels that are strictly isolated from each other, without mixing. We will next consider mixing due to spin-orbit interactions.

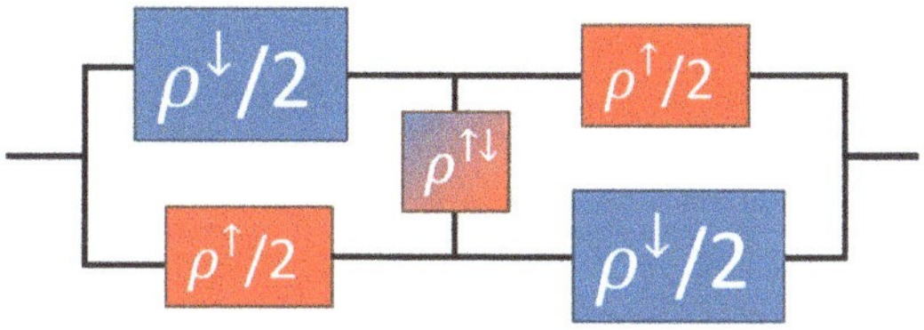

FIG. 16 – Illustration of how spin mixing, due to the spin-orbit interaction, alters the equivalent circuit representation of asymmetric electron-scattering, due to spin-dependent DOS.

At this stage, it is important to understand that the terminology 'mixing' means a mixing of eigenstates, where the electronic wavefunction becomes a linear combination of the $|\uparrow\rangle$ and $|\downarrow\rangle$ eigenstates – here due to spin-orbit interaction (see also §4.1 for a toy model describing mixing in the context of a transfer of angular momentum). We do insist on the fact that no spin-flip scattering needs to be considered at this stage. The spin-flip scattering will be tackled in chapter 3.

How spin mixing alters the equivalent circuit representation of asymmetric electron scattering is shown in figure 16 (*vs.* figure 10), where the resistivity $\rho^{\uparrow\downarrow}$ accounts for spin mixing. The equivalent resistivity of a ferromagnet is therefore given by

$$\rho = \frac{\rho^{\uparrow}\rho^{\downarrow} + \rho^{\uparrow\downarrow}(\rho^{\uparrow} + \rho^{\downarrow})}{\rho^{\uparrow} + \rho^{\downarrow} + 4\rho^{\uparrow\downarrow}} \tag{17}$$

Note that, in the absence of spin mixing, equation (17) naturally becomes equivalent to equation (9).

The term spin mixing refers interchangeably to spin and moment mixing.

The above spin mixing contribution $\rho^{\uparrow\downarrow}$ can be due to spin-orbit interaction, which couples the spin ($\boldsymbol{s}$) and orbital ($\boldsymbol{\ell}$) angular momenta of an electron, *i.e.*, the electron's spin and its motion. In *3d* transition metals, spin-orbit interaction causes two effects:

- **lifts the energy states** of the majority moment($\uparrow$)–spin($\Downarrow$) electrons and minority moment($\downarrow$)–spin($\Uparrow$) electrons, and

- **mixes or reorients the *d*-orbitals**. For *3d* electrons, with principal quantum number $n = 3$, the orbital quantum number and its projections along the quantization axis are $l = 2$ and $m_l = -2, -1, 0, +1, +2$ (figure 17).

In a quantum mechanical description, the corresponding Hamiltonian is

$$\boldsymbol{\mathcal{H}} = \xi_{SO}\widehat{\boldsymbol{L}}.\widehat{\boldsymbol{S}} \tag{18}$$

where ξ_{SO} is the spin-orbit coupling energy, with

$$\boldsymbol{L}.\boldsymbol{S} = L_x S_x + L_y S_y + L_z S_z = L_z S_z + \frac{1}{2}(\boldsymbol{L}^{+}\boldsymbol{S}^{-} + \boldsymbol{L}^{-}\boldsymbol{S}^{+}) \tag{19}$$

where $\boldsymbol{L}$ and $\boldsymbol{S}$ are the orbital and spin operators, and $\boldsymbol{L}^{+(-)}$ and $\boldsymbol{S}^{+(-)}$ are the corresponding ladder operators. These ladder operators involve the orbital and spin

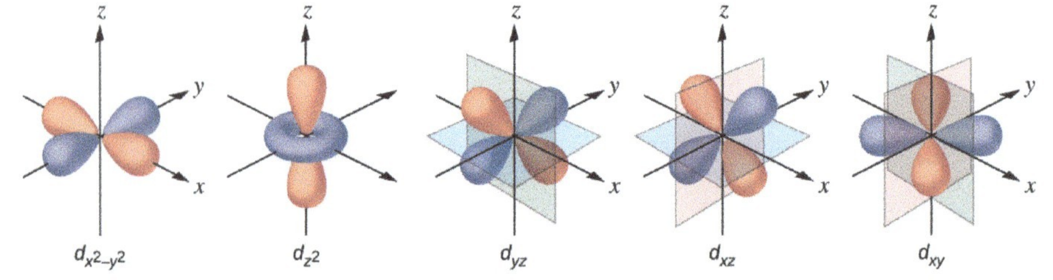

FIG. 17 – Electron probability densities for the *d*-orbitals in solid. The color-code refers to the sign of the wavefunction: blue for negative and red for positive. From P. Flowers *et al.*, *Chemistry*, OpenStax, Houston (2015) – https://openstax.org/books/chemistry/pages/1-introduction – CC BY License.

quantum numbers (l and s) and their respective projection quantum numbers (m_l and m_s) as follows

$$\begin{aligned} \boldsymbol{L}^{+(-)}|l, m_l\rangle &= \sqrt{l(l+1) - m_l(m_l + (-)1)}\hbar|l, m_l + (-)1\rangle \\ \boldsymbol{S}^{+(-)}|s, m_s\rangle &= \hbar|s, m_s + (-)1\rangle \end{aligned} \tag{20}$$

where $|l, m_l\rangle$ and $|s, m_s\rangle$ are the orbital and spin states.

Consequently, a $3d^{\uparrow(\downarrow)}(m_l)$ state will mix with a $3d^{\downarrow(\uparrow)}(m_l \pm 1)$ state by spin-orbit interaction. To set orders of magnitudes, ξ_{SO} can reach up $\sim 10\,\text{meV}$ in some *3d*-metals, to be compared with band-spittings of a few eV, and Fermi energies of a few tenth of eV ($1\,\text{eV} = 1.602 \times 10^{-19}\,\text{J}$).

A simplistic illustration of mixing of *d*-bands due to spin-orbit interactions considers that the majority moment(↑)–spin(⇓) electrons band acquires a minority moment(↓)–spin(⇑) character and *vice versa* (figure 18). This is a practical way to describe the fact that wavefunctions become a mix of eigenstates.

Based on the above description of spin mixing, we now describe the anisotropic magnetoresistance effect (AMR). For the sake of simplicity, we now consider a 'strong' ferromagnet. This term is used when the Fermi level crosses only one of the *3d*-bands (see Ni and Co in figure 7). It is opposed to the term 'weak' ferromagnet when both bands are crossed at the Fermi level (see Fe in figure 7). Thus for a strong ferromagnet, the majority moment(↑)–spin(⇓) electrons band is full. It is below the Fermi energy and no *d*-state is *a priori* allowed for scattering of a ↑ state: $N^{\uparrow}(\varepsilon_F) = 0$.

Without spin-orbit interaction, when $\xi_{SO} = 0$ (figure 19, left), the $s^{\uparrow}$ to $d^{\uparrow}$ scattering channel is shut due to strong ferromagnetism, and the following three scattering channels are open: $s^{\uparrow}$ to $s^{\uparrow}$, $s^{\downarrow}$ to $s^{\downarrow}$, and $s^{\downarrow}$ to $d^{\downarrow}$.

With spin-orbit interaction, when $\xi_{SO} \neq 0$ (figure 19, right), the mixing of *d*-bands modifies the spin-dependent scattering rates, mostly because it opens a new channel for scattering. Here, a $s^{\uparrow}$ to $d^{\uparrow}$ scattering channel opens up because the minority moment(↓)–spin(⇑) electrons band acquires a majority moment(↑)–spin(⇓) character. Consequently, the resistivity and the total resistance of the ferromagnet are changed, here it becomes larger.

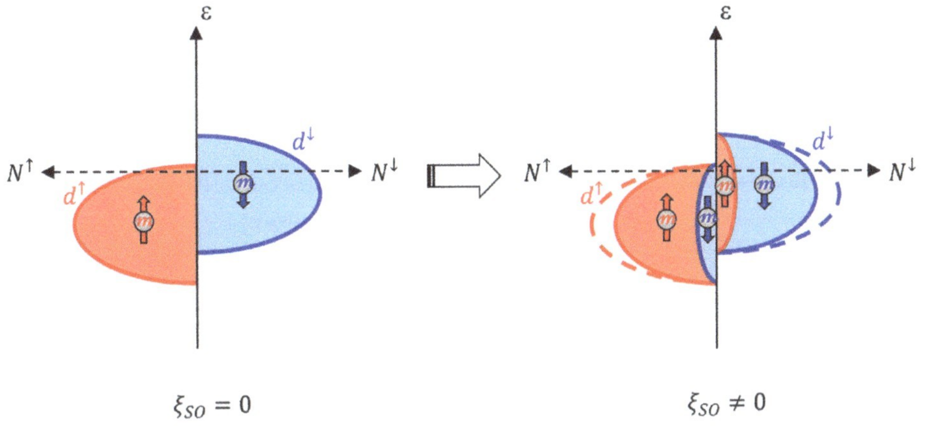

FIG. 18 – Illustration of how spin mixing, due to spin-orbit interaction (ξ_{SO}), alters the DOS of *d*-electrons. The illustration is simplified for the sake of clarity and does not mean that there is a loss of electrons. Adapted from S. Kokado *et al.*, *J. Phys. Soc. Jpn.* **81**, 024705 (2012) – CC BY license.

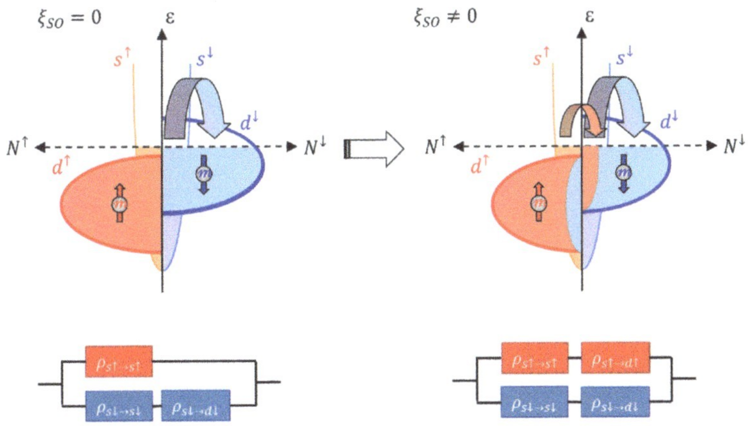

FIG. 19 – (Top) Illustration of how spin mixing, due to spin-orbit interaction (ξ_{SO}), alters *sd* electron scattering by opening a scattering channel – with the example of a strong ferromagnet, like Ni. (Bottom) Corresponding equivalent circuits.

A practical example of the effect of spin-orbit interaction on a ferromagnet's resistance is when the direction of an applied current $\boldsymbol{I} = \boldsymbol{J}_e A$, with A the cross-section surface, is varied with respect to the orientation of magnetization $\boldsymbol{M}$. The result of spin-orbit interactions, indeed, depends on the direction of the angular momentum of *s*-electrons (parallel to the wavevector $\hbar\boldsymbol{k}$) carrying the current relative to that of the *d*-electrons (parallel to $\boldsymbol{M}$) bearing magnetization. A simple picture is given in figure 20:

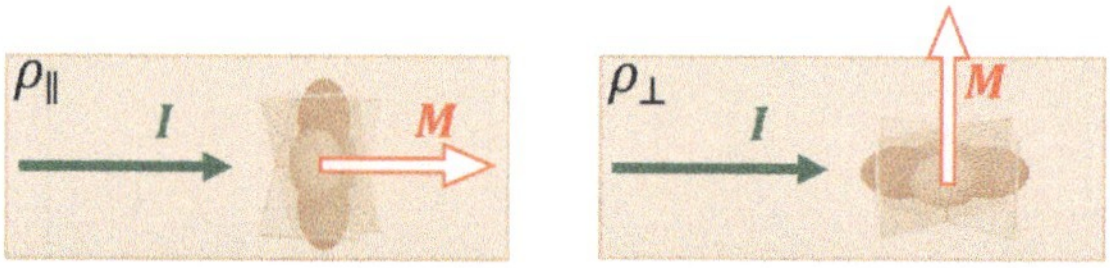

FIG. 20 – Representation of the anisotropic magnetoresistance (AMR) effect, which is a prototypical example of how spin mixing, due to spin-orbit interaction, influence electron transport. Here, spin mixing depends on the orientation between current (***I*** – linked to the angular momentum of *s*-electrons) and magnetization (***M*** – linked to the angular momentum of *d*-electrons). It is, for example, adjusted through the application of a magnetic field.

- **when** $\boldsymbol{I} \parallel \boldsymbol{M}$, the electronic orbits are $\perp$ to ***I***, offering a large cross-section for the *s*-electrons to scatter, resulting in a high resistivity $\rho_{\parallel}$. This is equivalent to the case depicted in figure 19, right.
- **when** $\boldsymbol{I} \perp \boldsymbol{M}$, the electronic orbits are $\parallel$ to ***I***, and thus offer a smaller cross-section for scattering, resulting in a low resistivity $\rho_{\perp}$. This is equivalent to the case depicted in figure 19, left.

This anisotropic magnetoresistance (AMR) effect was demonstrated experimentally by W. Thomson (Lord Kelvin) in 1857 and explained theoretically in the 1970ies by Campbell, Fert and Jaoul (CFJ model) [I. A. Campbell *et al.*, *J. Phys. C: Solid State Phys.* **3**, S95 (1970)]. AMR is expressed as follows

$$\frac{\Delta\rho}{\rho} = \frac{\rho_{\parallel} - \rho_{\perp}}{\rho_{\perp}} \tag{21}$$

Exercise 6.1 is devoted to AMR and, in particular, to the calculation of its angular dependence.

Some metals' typical resistivities and AMR ratios are given in figure 21 (see also figure 14 for comparison with GMR magnitudes). The composition-dependence of

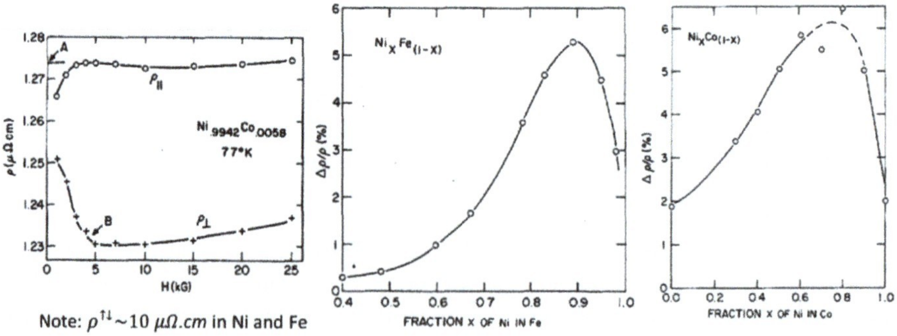

FIG. 21 – Experimentally determined values of AMR for several compositions of NiCo and NiFe alloys. Reprinted with permission from IEEE: T. McGuire and R. Potter, *IEEE Trans. Magn.* **11**, 1018 (1975). Copyright 1975.

AMR partly relates to band filling, *i.e.*, to how the Fermi level is moved with respect to the densities of states, which will alter scattering probability.

Note that, the presence of a magnetic texture, like a domain wall (DW), may change the relative orientation between current and magnetization compared to the uniform state, and subsequently alter the resistivity of the system. This effect is known as the DWAMR and it will be evoked when discussing the general context of DWMR [equation (43) and corresponding text]. In addition, exercise 6.2 is devoted to DWAMR.

Summary

In this chapter, we have seen that spin-dependent transport in band ferromagnets are due to an asymmetric scattering probability due to spin-splitting of the majority moment($\uparrow$)–spin($\Downarrow$) electrons band relative to the minority moment($\downarrow$)–spin($\Uparrow$) electrons band, $N^{\uparrow}(\varepsilon_F) \neq N^{\downarrow}(\varepsilon_F)$. This asymmetry leads to a spin-dependent mean free path $\lambda_e^{\uparrow(\downarrow)}$ and related Ohm's law $\boldsymbol{j}^{\uparrow(\downarrow)} = \sigma^{\uparrow(\downarrow)}\boldsymbol{\nabla}\mu^{\uparrow(\downarrow)}/e$. It is formulated in terms of the basic spin asymmetry parameter $\alpha = \rho^{\downarrow}/\rho^{\uparrow}$, and is at the origin of the CIP-GMR effect. Adding spin-orbit interactions to the problem generates spin mixing $\rho^{\uparrow\downarrow}$ which can be thought of as a mixing of the majority moment($\uparrow$)–spin($\Downarrow$) and minority moment($\downarrow$)–spin($\Uparrow$) electrons bands, therefore reshuffling the spin-dependent scattering probabilities. The spin-orbit-related spin mixing mechanism is at the origin of the AMR effect.

Chapter 3

Spin Accumulation – CPP-GMR

In chapter 2, we have seen that longitudinal transport is spin-dependent in ferromagnets. It is formulated in terms of the basic asymmetry parameter $\alpha = \rho^{\downarrow}/\rho^{\uparrow}$, which is due to distinct spin-dependent DOS at the Fermi level, leading to a spin-dependent mean free path, and thus causing the current density to be different for the majority moment(↑)–spin(⇓) electrons and minority moment(↓)–spin(⇑) electrons: $\boldsymbol{j}^{\uparrow} \neq \boldsymbol{j}^{\downarrow}$. As a result, when electrons flow across an interface (figure 22), the distinct partial densities of current at the interface between materials of different types with different asymmetry parameters, for example, a ferromagnet and a non-magnet create a spin imbalance. This effect is called spin accumulation.

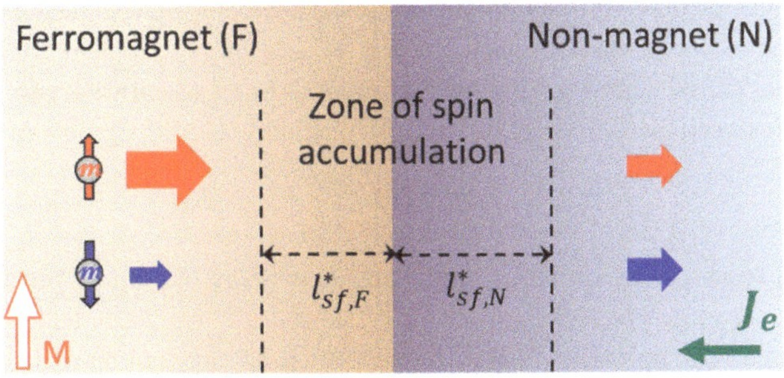

FIG. 22 – Schematics of a ferromagnet/non-magnet (F/N) metallic bilayer subjected to a perpendicular-to-plane charge current, illustrating the creation of a zone of spin accumulation at the interface, which accommodates for the difference in spin asymmetry (§2.4) between the two layers.

How electron dynamics is described in such a case; which types of relaxation towards the steady state regime prevail; which conditions are necessary to maintain spin polarization across the interface; and how spin accumulation is used in spintronic devices, such as CPP-GMR sensors, will be detailed in the next sections.

DOI: 10.1051/978-2-7598-2917-0.c003

3.1 Spin-Dependent Spin-Flip Scattering

We first introduce the spin-flip scattering relaxation process (figure 23), by which the spin is eventually flipped upon scattering, after about a hundred non-spin-flip events. Spin-flip scattering makes the majority moment(↑)–spin(⇓) electrons channel communicate with the minority moment(↓)–spin(⇑) electrons channel and *vice-versa*.

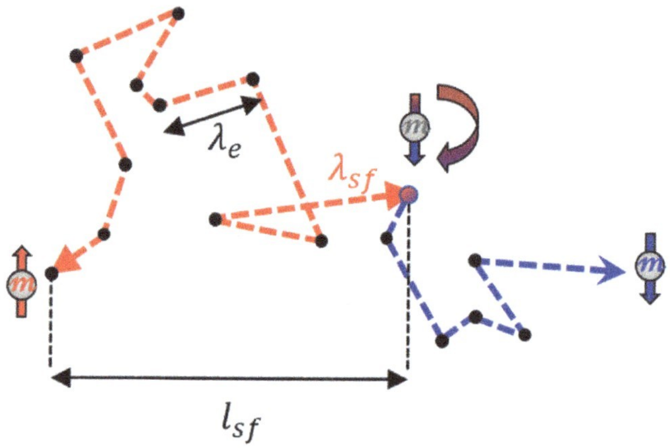

FIG. 23 – Representation of electron scattering when a spin-flip scattering event is considered. λ_e is the (spin-dependent) electron mean free path. λ_{sf} is the (spin-dependent) spin-flip length (total length of the red path). l_{sf} is the (spin-dependent) spin-diffusion length; it is in a way λ_{sf} reduced to a given direction, which is set by an applied current in a real three-dimensional experimental setting. Note: to make the representation easy to read, λ_e and λ_{sf} are shown between particular collisions. In practice, λ_e and λ_{sf} are averaged over all collisions.

In addition to the electron mean free path $\lambda_e = v_F \tau_e$ (§2.2), a spin-flip length can be defined $\lambda_{sf} = v_F \tau_{sf}$. To avoid misleading the reader, we want to insist on the fact that, in figure 23, the spin-flip length corresponds to the length of the path colored in red. What matters for CPP transport, is the direction perpendicular to the interface. λ_{sf} is thus barely used and another length called the spin diffusion length l_{sf}, is more relevant. Indeed, it describes the length between two spin-flip scattering events, along a given direction. Based on Einstein's relation, we have $l_{sf} = \sqrt{D\tau_{sf}}$. Based on Fick's law, the diffusion constant D is written $D = \frac{1}{3} v_F \lambda_e$ and the spin-diffusion length can also be expressed as $l_{sf} = \sqrt{\lambda_e \lambda_{sf}/3}$. This expression underlines that l_{sf} is the geometrical mean between λ_e and λ_{sf}.

Recall that $l_{sf} > \lambda_e$, which has consequences for the GMR in CIP *vs.* CPP geometries [§2.5, figure 25 and J. Bass and W. P. Pratt Jr, *J. Phys.: Condens. Matter* **19**, 183201 (2007) for numbers]. In some cases, spin-flip events can be as scarce as 0.1%. It is also important to note that, because the mean free path is spin-dependent, the spin-diffusion length will naturally be spin-dependent. In addition, we admit here that the spin-diffusion length is also spin-dependent because spin-flip events can be spin-dependent. The spin-dependent spin-diffusion length therefore reads $l_{sf}^{\uparrow(\downarrow)} = \sqrt{D^{\uparrow(\downarrow)}\tau_{sf}^{\uparrow(\downarrow)}}$ (two equations).

We point out that, $\tau_{sf}^{\uparrow}$ and $\tau_{sf}^{\downarrow}$ represent the time for a majority moment($\uparrow$)–spin ($\Downarrow$) electron and a minority moment($\downarrow$)–spin($\Uparrow$) electron to be flipped, respectively. The notations $\tau^{\uparrow\downarrow}$ and $\tau^{\downarrow\uparrow}$ are sometimes used in the literature for spin-flip scattering. However, these latter notations can be misleading. In this book, the symbol $\uparrow\downarrow$ (page 151) is exclusively used to describe spin-mixing processes, *i.e.*, mixing of the eigenstates, in the absence of spin-flipping (§2.6).

3.2 Drift-Diffusion Model, Spin-Coupled Resistance & Mismatch

In the previous sections, we introduced the idea that the distinct partial densities of current induced at the interface between two materials of different natures create a spin accumulation (figure 22). Relaxation towards equilibrium conditions causes spins to diffuse near the interface. The *average* spin-diffusion lengths on either side of the interface $l^*_{sf,F(N)}$ are the characteristic lengths for this relaxation process. How $l^*_{sf,F(N)}$ come into play and how they connect to $l^{\uparrow(\downarrow)}_{sf,F(N)}$ will be detailed in this section [equation (27)]. We remind the reader that the notation $l^*_{sf,F(N)}$ accounts for two lengths: $l^*_{sf,F}$ in the F layer and $l^*_{sf,N}$ in the N layer; and that $l^{\uparrow(\downarrow)}_{sf,F(N)}$ thus accounts for four lengths: $l^{\uparrow}_{sf,F}$, $l^{\downarrow}_{sf,F}$, $l^{\uparrow}_{sf,N}$, and $l^{\downarrow}_{sf,N}$.

The purpose of what follows is to determine, in the whole F/N stack:

- **the spin $\boldsymbol{s}$ accumulation** or chemical potential imbalance: $\mu_s = -(\mu^{\uparrow} - \mu^{\downarrow})$,
- **the spin $\boldsymbol{s}$ current**: $\boldsymbol{J}_s = -(\boldsymbol{j}^{\uparrow} - \boldsymbol{j}^{\downarrow})$.

Next, we show all the steps that allow us to find the expressions of μ_s and $\boldsymbol{J}_s$ as a function of the layers' parameters. At this stage, we skip the calculations, in order to go straight to a physical understanding of the phenomenon. Detailed calculations are given in exercise 6.3 and in P. C. von Son *et al.*, *Phys. Rev. Lett.* **58**, 2271 (1987), and T. Valet and A. Fert, *Phys. Rev. B* **48**, 7099 (1993).

In the literature, spin accumulation is defined either as $\mu_s = \mu^{\uparrow} - \mu^{\downarrow}$, or $\mu_s = -(\mu^{\uparrow} - \mu^{\downarrow})$, or $\mu_s = \pm(\mu^{\uparrow} - \mu^{\downarrow})/2$, resulting in a factor $\pm 1/2$ difference in some of the equations, depending on the articles and books consulted. Throughout this book, the term spin refers to $\boldsymbol{s}$ and the spin $\boldsymbol{s}$ accumulation is thus defined as $\mu_s = \mu^{\Uparrow} - \mu^{\Downarrow} = -(\mu^{\uparrow} - \mu^{\downarrow})$, because the $\uparrow$ and $\downarrow$ symbols refer to the orientation of the moment $\boldsymbol{m}$, whereas the $\Uparrow$ and $\Downarrow$ symbols refer to the orientation of the spin $\boldsymbol{s}$, which points in the opposite direction to $\boldsymbol{m}$ (see symbols and units page 150).

■ **Step 1: generalized Ohm's law**

We first recall the generalized Ohm's law (4 equations)

$$j_{F(N)}^{\uparrow(\downarrow)} = \sigma_{F(N)}^{\uparrow(\downarrow)} \frac{\nabla \mu_{F(N)}^{\uparrow(\downarrow)}}{e} \tag{22}$$

with $\mu_{F(N)}^{\uparrow(\downarrow)} = \mu_{n,F(N)}^{\uparrow(\downarrow)} = \mu_{n_0} + \frac{e^2 D_{F(N)}^{\uparrow(\downarrow)}}{\sigma_{F(N)}^{\uparrow(\downarrow)}} \delta n_{F(N)}^{\uparrow(\downarrow)}$. Drift, described by the electrostatic potential μ_e (§2.2), is omitted here to facilitate reading. It will cause straightforward contributions to add up. In fact, in the most general case, μ is defined as $\mu = \mu_e + \mu_n$ and is called the electrochemical potential. μ_n is called the chemical potential (linked to the diffusion process). It corresponds to the energy necessary to add one particle (here, electrons) to the system and thus equals the Fermi energy at equilibrium and zero temperature. It is defined as $\mu_n = \mu_{n_0} + \frac{\delta n}{N(\varepsilon_F)} = \mu_{n_0} + \frac{e^2 D}{\sigma} \delta n$, where μ_{n_0} is the equilibrium chemical potential and δn is the electron density in excess compared to the steady state in the isolated layer.

It is important to note that μ_n is spin-dependent because σ and δn are spin-dependent parameters. It is also important to understand that, in the absence of spin-accumulation, the excess particle density as defined above is equal to zero, and $\mu_n = \mu_{n_0}$. So, in a single layer at steady-state equilibrium, $\mu = \mu_e + \mu_{n_0}$ and $\nabla \mu = \nabla \mu_e$, as used in chapter 2, see *e.g.* equation (7).

Obtaining equation (22) and the equivalence to spin-dependent macroscopic transport equations is not as straightforward as it may seem. The accurate calculation uses Boltzmann transport theory. How the Boltzmann equation model reduces to macroscopic transport equations when $l_{sf}^* \gg \lambda_e$ is extensively described by T. Valet and A. Fert in *Phys. Rev. B* **48**, 7099 (1993).

In the following, the F and N subscripts will be omitted to facilitate reading. The equations and steps discussed below, apply in the F and in the N layer separately.

■ **Step 2: charge and spin currents**

Recall that the definitions of the charge and spin current densities [equations (3) and (4)] are

$$\boldsymbol{J}_e = \boldsymbol{j}^{\uparrow} + \boldsymbol{j}^{\downarrow} \tag{23}$$

$$\boldsymbol{J}_s = -(\boldsymbol{j}^{\uparrow} - \boldsymbol{j}^{\downarrow}) \tag{24}$$

■ **Step 3: charge conservation**

The flux of charge current density is conservative

$$\nabla \cdot \boldsymbol{J}_e = 0 \tag{25}$$

■ Step 4: total angular momentum conservation

The flux of angular momentum current density $\boldsymbol{Q} = \frac{\hbar}{2e}\boldsymbol{J}_s$ is not conservative. However, total angular momentum is conserved. As stated by Lavoisier: "Nothing is lost, nothing is created, everything is transformed", meaning that any loss in spin current must be due to spin-flips (at this stage). This is expressed as follows

$$\boldsymbol{\nabla} \cdot \boldsymbol{J}_s = -e\left(\frac{\delta n^{\uparrow}}{\tau^{\uparrow}_{sf}} - \frac{\delta n^{\downarrow}}{\tau^{\downarrow}_{sf}}\right) \tag{26}$$

At this stage, we only consider spin-flip scattering as the process by which spin current is lost. There are several other sources of spin angular momentum relaxation, *e.g.*, electron–phonon, electron–magnon, and electron–electron interactions, besides the spin-flip scattering process discussed here, but also the spin mixing process discussed in §2.4 and *sd* transfer of angular momentum through the spin torques described in chapter 4.

■ Step 5: obtain the diffusion equation for spin accumulation

Using equations (22)–(26), it is possible to show (exercise 6.3) that CPP transport at a heterogeneous interface is governed by a simple diffusion equation describing spin accumulation (or chemical potential imbalance) $\mu_s = -(\mu^{\uparrow} - \mu^{\downarrow})$

$$\boldsymbol{\nabla}^2 \mu_s - \frac{\mu_s}{l^{*\,2}_{sf}} = 0 \tag{27}$$

where l^*_{sf} can be viewed as an *average* spin-diffusion length, because $\frac{1}{l^{*\,2}_{sf}} = \frac{1}{l^{\uparrow\,2}_{sf}} + \frac{1}{l^{\downarrow\,2}_{sf}}$. In the literature, l^*_{sf} is most often written as l_{sf}. The notation l_{sf} may implicitly infer that the spin-diffusion length is spin-independent, which is not always the case. Hence, we prefer to use a superscript as a warning.

When a time-dependent chemical potential $\frac{\partial \mu_s}{\partial t}$ (§6.6) and external magnetic field $\boldsymbol{H}$ come into play, it is possible to show that equation (27) becomes

$$\frac{\partial \boldsymbol{\mu}_s}{\partial t} = -|\gamma|\boldsymbol{\mu}_s \times (\mu_0 \boldsymbol{H}) + D\boldsymbol{\nabla}^2 \boldsymbol{\mu}_s - \frac{\boldsymbol{\mu}_s}{\tau_{sf}} \tag{28}$$

where $\boldsymbol{\mu}_s$ is a vector (§3.5 and §6.6) and $-|\gamma|\boldsymbol{\mu}_s \times (\mu_0 \boldsymbol{H})$ accounts for the possibility of spin accumulation to precess around the magnetic field. The gyromagnetic ratio is $\gamma = -\frac{e}{2m_e} g$, and the Landé *g*-factor is $g \sim 2$ for an isolated electron, making $\frac{|\gamma|}{2\pi} \sim 28\,\text{GHz.T}^{-1}$.

Other effects may influence spin accumulation. An example of how the spin-dependent Seebeck effect (§5.5) alters the drift-diffusion equation for spin accumulation is shown in exercise 6.3.

■ Step 6: solve the diffusion equation for spin accumulation

In a one-dimensional problem, in which the current flows along x, the solutions of equation (27) take the form (two equations)

$$\mu_{s,F(N)}(x) = A_{F(N)} e^{x/l^*_{sf,F(N)}} + B_{F(N)} e^{-x/l^*_{sf,F(N)}} \tag{29}$$

where $A_{F(N)}$ and $B_{F(N)}$ are four constants, obtained from the boundary conditions.

For infinite layers, we have

$$\lim_{x \to -(+)\infty} \left(\mu_{s,F(N)} \right) = 0 \tag{30}$$

We note that exercise 6.6 proposes a problem in which the diffusion equation for spin accumulation is solved for finite-size layers.

Continuity of μ_s (due to continuity of $\mu^{\uparrow(\downarrow)}$) and J_s across the interface gives

$$\mu_{s,F}(0) = \mu_{s,N}(0) \tag{31}$$

$$J_{s,F}(0) = J_{s,N}(0) \tag{32}$$

Using equations (27)–(32), it is possible to calculate the spin accumulation in the F/N bilayer (figure 24, see T. Valet and A. Fert, *Phys. Rev. B* **48**, 7099 (1993) for detailed calculations)

$$\mu_{s,F(N)}(x) = 2e\beta J_e l^*_{sf,N} \rho_N \left(1 + \frac{l^*_{sf,N} \rho_N}{l^*_{sf,F} \rho^*_F} \right)^{-1} e^{+(-)\frac{x}{l^*_{sf,F(N)}}} \tag{33}$$

with $\rho^*_F = \rho_F / (1 - \beta^2)$. We recall that $\beta = \frac{\alpha - 1}{\alpha + 1}$ is the polarization of the current in the F layer, which is linked to the spin asymmetry parameter $\alpha = \frac{\rho^\downarrow_F}{\rho^\uparrow_F}$ due to the spin-dependent mean free path (§2.4).

From figure 24, middle and equation (33), we see that the *average* spin-diffusion lengths on either side of the interface $l^*_{sf,F(N)}$ are the characteristic lengths for spin accumulation, as anticipated in the introduction of this chapter (see also figure 25 for numbers).

Note that $\mu_{s,F(N)}$ is proportional to J_e, meaning that increasing the current in a device will increase spin accumulation, though at the expanse of Joule heating and power consumption. In addition, because the spin accumulation is proportional to the charge current, it is odd in J_e and consequently changes signs under reversal of the current direction, $\mu_{s,F(N)}(J_e) = -\mu_{s,F(N)}(-J_e)$. This also implies that the spin accumulation changes sign under the reversal of the stacking order, $\mu_{s,F(N)}(F/N) = -\mu_{s,F(N)}(N/F)$. These properties will come into play in the next section (§3.3), when discussing F/N/F heterostructures in P and AP states. Typical values of charge currents used to estimate the effects related to spin accumulation are: $J_e \sim 10^5 - 10^6$ A.cm^{-2}, and typical magnitudes of spin accumulation are $\mu_s \sim 10 - 100$ μeV.

Note also that, $\mu_{s,F(N)}$ is proportional to β, reminding us that efficient spin accumulation is obtained by selecting appropriate materials, especially by selecting a F layer with a large spin asymmetry parameter, α (see figure 25 for numbers). Spin accumulation is thus odd in β and then changes sign under magnetization reversal,

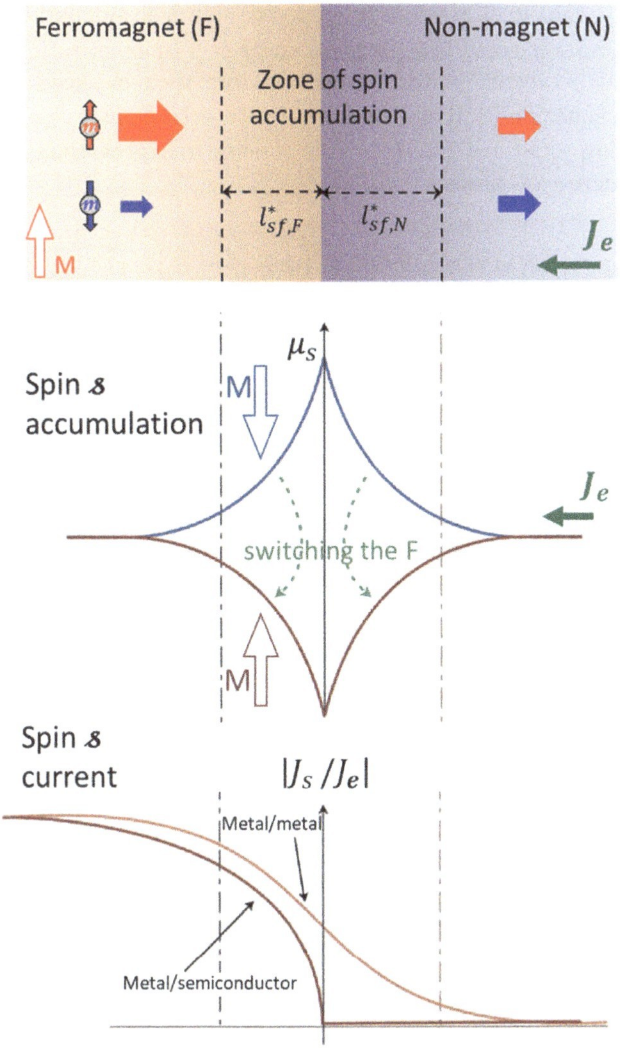

FIG. 24 – (Top) Schematics of a ferromagnet/non-magnet bilayer, illustrating the zone of spin accumulation at the interface. (Middle) Corresponding spin s accumulation $\mu_s = -(\mu^{\uparrow} - \mu^{\downarrow})$, calculated for two orientations of magnetization $\boldsymbol{M}$. The ↑ and ↓ symbols refer to the orientation of the moment $\boldsymbol{m}$. Drift is omitted to facilitate understanding. (Bottom) Corresponding ratio between spin s current $\boldsymbol{J}_s = -(\boldsymbol{j}^{\uparrow} - \boldsymbol{j}^{\downarrow})$ and charge current $\boldsymbol{J}_e = \boldsymbol{j}^{\uparrow} + \boldsymbol{j}^{\downarrow}$, calculated for the case of metallic and semiconducting non-magnets. Adapted with permission from EDP Sciences: A. Fert, *Reflets Phys.* **15**, 5 (2009).

as illustrated in figure 24, middle, $\mu_{s,F(N)}(M) = -\mu_{s,F(N)}(-M)$, or written differently to ease further understanding, $\mu_{s,F(N)}(F^{\uparrow}) = -\mu_{s,F(N)}(F^{\downarrow})$. This property will also come into play in §3.3.

Finally, note that $\mu_{s,F(N)}$, which refers to the spin $\boldsymbol{s}$ accumulation, is negative for $J_e<0$ (charge current going from N to F) and $\beta > 0$ (magnetization 'pointing up'), as the majority moment(↑)–spin(⇓) electrons are then in excess at the interface [figure 24 and equation (33)] and $\mu_s = \mu^{\Uparrow} - \mu^{\Downarrow} = -(\mu^{\uparrow} - \mu^{\downarrow})$.

Further using equations (22)–(32), it is possible to express the spin current in the F/N bilayer (figure 24, bottom)

$$J_{s,F}(x) = -\beta J_e \left[1 - \left(1 + \frac{l^*_{sf,F}\rho^*_F}{l^*_{sf,N}\rho_N}\right)^{-1} e^{\frac{x}{l^*_{sf,F}}}\right] \tag{34}$$

$$J_{s,N}(x) = -\beta J_e \left(1 + \frac{l^*_{sf,N}\rho_N}{l^*_{sf,F}\rho^*_F}\right)^{-1} e^{\frac{-x}{l^*_{sf,N}}} \tag{35}$$

From figure 24, bottom and equations (34) and (35), it becomes clear that, as the result of spin accumulation, spin polarization diffuses from the F layer to the N layer. It is also said that spin is injected in the N layer. The same remarks as above can be drawn on the odd symmetry of $J_{s,F(N)}$ with J_e and β.

	Bulk resistivity $\rho[\mu\Omega\,\text{cm}]$	Bulk asymmetry β	Spin diffusion length $l^*_{sf}[\text{nm}]$
Co	15	0.35	38
Cu	2	0	100
Ru	20	0	14
$Co_{90}Fe_{10}$	20	0.7	25
$Ni_{84}Fe_{16}$	20	0.6	4.5
$Cu_{94}Pt_6$	35	0	7
	Interfacial resistance $r[\text{m}\Omega\,\mu\text{m}^2]$	Interfacial spin asymmetry γ	Interfacial spin memory loss δ
Co/Cu	45	0.7	25%
Co/Ru	48	−0.2	
$Co_{90}Fe_{10}$/Cu	45	0.75	25%
$Co_{90}Fe_{10}$/Ru	48	−0.2	31%
Cu/Ru	48	0	35%
$Ni_{50}Fe_{50}$/Cu	25	0.63	33%

FIG. 25 – Examples of the bulk and interface (§3.4) parameters at stake in spintronic effects involving spin accumulation. Note: r, γ and δ are parameters defined in the next paragraph. They take into account spin-dependent process due to the interface itself. Reprinted with permission from the American Physical Society: A. Manchon *et al.*, *Phys. Rev. B* **73**, 184418 (2006). Copyright 2006. See also J. Bass and W. P. Pratt Jr, *J. Phys.: Condens. Matter* **19**, 183201 (2007) for more numbers.

■ Spin resistance

In the same manner, as ρ (σ) is called the resistivity (conductivity) because it accounts for any process that opposes (favors) the flow of electrons, ρl^*_{sf} (σ/l^*_{sf}) is naturally called the spin resistance/impedance (spin conductance) as it opposes (favors) the flow of spins. This notion is commonly encountered in the field of spintronics, especially when dealing with spin injection and transmission, as we are here. Note however that, it is customary to incorrectly use the term spin 'resistance' and 'conductance', although these quantities refer to a resistance area product in $\Omega.m^2$ and a conductance per area in $S.m^{-2}$, respectively. In practice, spin conductance is often used per quantum conductance $\frac{\sigma/l^*_{sf}}{G_0}$, in m^{-2}. It can be compared directly to the spin mixing 'conductance' $g^{\uparrow\downarrow} = \frac{G^{\uparrow\downarrow}}{AG_0/2}$ (§4.2).

■ Spin impedance mismatch

To further underline the importance of the choice of the layers' parameters for efficient spin injection, we remark from equations (34) and (35) that, when $l^*_{sf,N}\rho_N \gg l^*_{sf,F}\rho^*_F$, for example, in the case of an F-metal/N-semiconductor bilayer (figure 24, bottom), spins accumulate and depolarize in the F layer. No spin polarization diffuses in the N layer, as $J_{s,N} = 0$. This issue is known as the spin impedance mismatch. Overcoming this matter requires adding a large spin-dependent resistance at the interface, *e.g.*, a tunnel barrier. More about impedance mismatch is explained in exercise 6.4, based on A. Fert and H. Jaffrès, *Phys. Rev. B* **64**, 184420 (2001).

■ Discontinuity of the potential and corresponding spin-coupled interface resistance

As a final remark on this part, we discuss the notion of a discontinuity of the potential at the interface, often encountered in papers. First, to obtain the electrochemical potential $\mu^{\uparrow(\downarrow)}_{F(N)}$ the (drift) contribution of the electrostatic potential $\mu_{e,F(N)}$, which we initially omitted to facilitate reading (see step 1, page 34), shall be added to the chemical potential $\mu^{\uparrow(\downarrow)}_{n,F(N)}$. From equation (7) and the corresponding text, it naturally follows that the electronic potential is linear with x: $\mu_{e,F(N)}(x) = eJ_e\rho_{F(N)}x$. The plots of the electrochemical potentials for majority moment($\uparrow$)–spin($\Downarrow$) electrons and minority moment($\downarrow$)–spin($\Uparrow$) electrons are given in figure 26.

We note that the out-of-equilibrium chemical contributions $\mu^{\uparrow(\downarrow)}_{n,F(N)}$ to the electrochemical potentials $\mu^{\uparrow(\downarrow)}_{F(N)} = \mu_{e,F(N)} + \mu^{\uparrow(\downarrow)}_{n,F(N)}$ vanish far from the interface. Indeed, in this case, the F and N layers behave as if they were isolated, when there is no excess density δn of majority moment($\uparrow$)–spin($\Downarrow$) electrons and minority moment ($\downarrow$)–spin($\Uparrow$) electrons.

From figure 26, a simplification consists in considering only the linear dependence of the electrostatic potentials $\mu_{e,F(N)}$ on either side of the interface, and how they mismatch at the interface ($\Delta\mu_e$) when extrapolated back there. In fact, this mismatch is due to the non-monotonic contributions of the chemical potentials $\mu^{\uparrow(\downarrow)}_{n,F(N)}$ subsequent to spin accumulation. These latter continuous contributions of

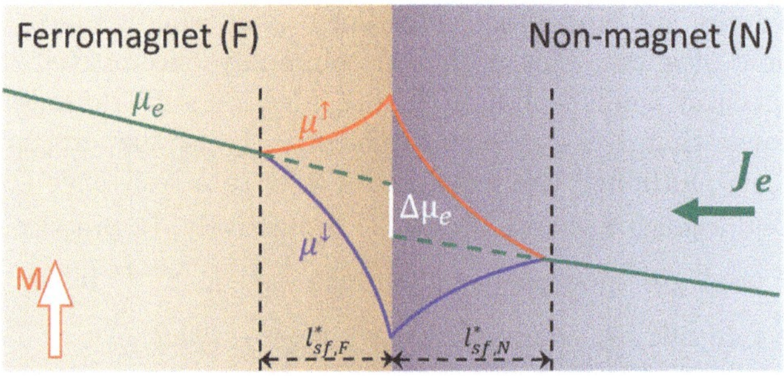

FIG. 26 – Illustrations showing how spin accumulation $\mu_s = -(\mu^\uparrow - \mu^\downarrow)$ can be reduced to a discontinuity $-\Delta\mu_e$ of the electronic potential, and henceforth, to a simple spin-coupled interface resistance area product $r_s = -\Delta\mu_e/(eJ_e)$ and corresponding voltage drop.

$\mu^{\uparrow(\downarrow)}_{n,F(N)}$ [equation (31)] which extend over finite sizes defined by the spin diffusion lengths $l^*_{sf,F(N)}$ are thus relegated to a discontinuity of electrostatic potential $\Delta\mu_e$ and subsequently to a simple voltage drop. It can be expressed in terms of a spin-coupled interface 'resistance' (in fact a resistance area product in $\Omega.\text{m}^2$)

$$r_s = \frac{-\Delta\mu_e}{eJ_e} = \frac{\mu_{s,F}(0)}{eJ_e} = \frac{\mu_{s,N}(0)}{eJ_e} \tag{36}$$

Using equation (33), one obtains

$$r_s = 2\beta l^*_{sf,N}\rho_N\left(1 + \frac{l^*_{sf,N}\rho_N}{l^*_{sf,F}\rho^*_F}\right)^{-1} \tag{37}$$

This spin-coupled interface resistance can also be reworked as follows

$$r_s = \beta^* l^*_{sf,N}\rho_N \tag{38}$$

where β^* is the effective spin polarization at the interface.

3.3 Heterostructures – CPP-GMR

In the previous section, we studied how a CPP current creates spin accumulation at an F/N interface. We discussed which parameters are essential – considering spin-flip scattering only – and calculated the thickness-dependence of spin-accumulation, as driven by a macroscopic (drift-)diffusion equation. Let us now consider the case of a CPP current in an F/N/F trilayer (figure 27), in which spin accumulation: (i) occurs at the F/N and N/F interfaces and (ii) depends on the relative orientation of the F layers.

In the following, when commenting on figure 27, we will focus on the central part consisting of the F/N/F trilayer and discard the outside parts, for which periodic boundary conditions were set in order to ease calculations.

In the $F^{\uparrow}/N/F^{\uparrow}$ P state (figure 27, left), for $J_e > 0$, the spin accumulation μ_s is positive at the $F^{\uparrow}/N$ interface and negative at the $N/F^{\uparrow}$ interface. It is a consequence of the odd symmetry of μ_s with stacking order: $\mu_{s,F(N)}(F^{\uparrow}/N) = -\mu_{s,F(N)}(N/F^{\uparrow})$ (page 36).

Conversely, in the $F^{\uparrow}/N/F^{\downarrow}$ AP state (figure 27, right), the spin accumulation μ_s is positive at the $F^{\uparrow}/N$ interface and positive at the $N/F^{\downarrow}$ interface. It is a consequence of the odd symmetry of μ_s with stacking order combined with the odd symmetry of μ_s with magnetization reversal (page 36): $\mu_{s,F(N)}(F^{\uparrow}/N) = -(-)\mu_{s,F(N)}(N/F^{\downarrow}) = \mu_{s,F(N)}(N/F^{\downarrow})$.

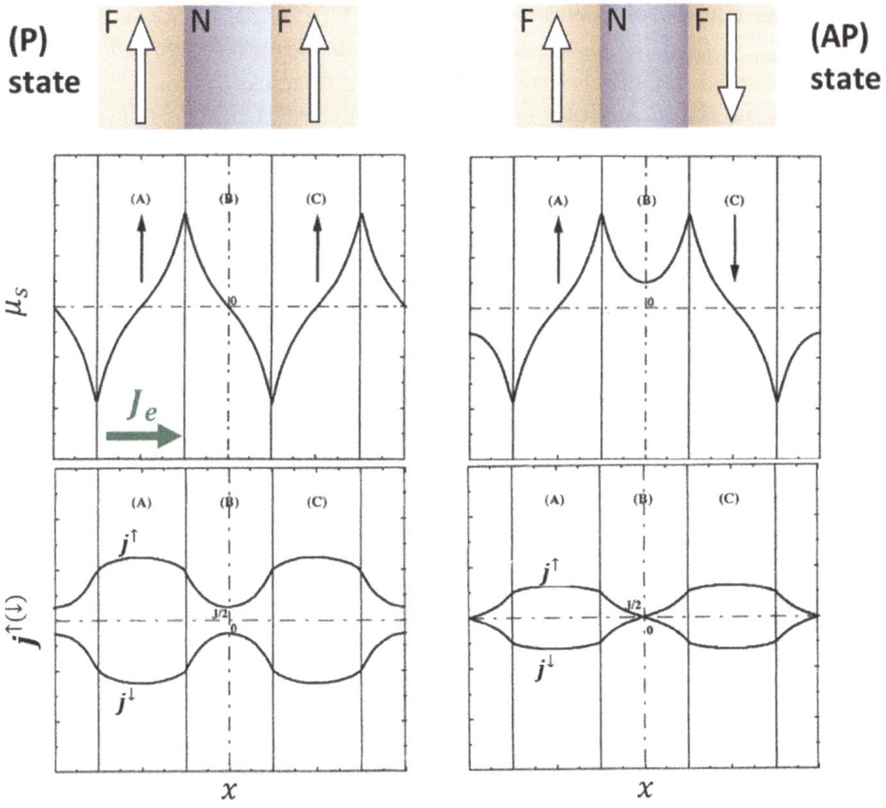

FIG. 27 – (Top) Schematics of a ferromagnet/non-magnet/ferromagnet (F/N/F) trilayer set in the parallel (P) and antiparallel (AP) magnetic states. (Middle) Corresponding calculation of the spin $\boldsymbol{s}$ accumulation μ_s. Drift is omitted to facilitate understanding. (Bottom) Corresponding charge currents for majority moment(↑)–spin(⇓) electrons and minority moment (↓)–spin(⇑) electrons. Periodic conditions are set outside the F/N/F trilayer, to ease calculations. Adapted with permission from the American Physical Society: T. Valet and A. Fert, *Phys. Rev. B* **48**, 7099 (1993). Copyright 1993.

Because spin accumulation in the P and AP states differs, the resulting spin-coupled resistances at the interfaces also differ and so does the overall resistance of the stack. For small enough thicknesses d_F and d_N of the F and N layer, respectively, such that $d_{F(N)} \ll l^*_{sf,F(N)}$, it is possible to show that the expression for CPP-GMR reads

$$\text{CPP-GMR} = \frac{\rho^{AP} - \rho^{P}}{\rho^{AP}} = \beta^2 \frac{\left(2\rho^*_F d_F\right)^2}{\left(\rho^*_N d_N + 2\rho^*_F d_F\right)^2} \tag{39}$$

The exact derivation of CPP-GMR can for example be found in *Phys. Rev. B* **48**, 7099 (1993). Note that CPP-GMR vanishes for $d_{F(N)} \gg l^*_{sf,F(N)}$, especially because the F/N and N/F interfaces become independent and disconnected when $d_N \gg l^*_{sf,N}$.

Examples of how typical parameters influence CPP-GMR are given in figure 28. Because $l^*_{sf,\text{Au}} \ll l^*_{sf,\text{Cu}}$ it follows that CPP-GMR falls much faster in CoFe(1)/Au (d_N)/CoFe(3) than in CoFe(1)/Au(d_N)/CoFe(3) (nm) stacks [figure 28c, left]. When d_F increases in NiFe(d_F)/CoFe(1)/Cu(4)/CoFe(3) and CoFe(d_F)/Cu(4)/CoFe(3) (nm) stacks, the absolute MR increases, as more polarization builds up, and it then levels out when $d_F \sim l^*_{sf,F}$ [figure 28b, right]. Above the threshold thickness $d_F \sim l^*_{sf,F}$, adding more film makes no further contribution to polarization, but it increases the CPP resistance instead, explaining the non-monotonic dependence of CPP-GMR with d_F [figure 28c, right].

3.4 Interface Scattering, Spin Memory Boost, and Spin Memory Loss

In the previous sections on CPP-GMR, bulk spin-dependent scattering processes were considered; through the bulk spin-dependent mean free path $\lambda^{\uparrow(\downarrow)}_{e,F(N)}$, and spin-flip scattering processes *via* $l^*_{sf,F(N)}$. When the resulting spin accumulation can be modelled by a spin-coupled interface resistance, this type of resistance is not related to the spin-dependent properties that are inherent to the interface itself.

Refined models must consider interfaces as well, because, for example, decoherent roughness-induced stray fields or changes in the local DOS near interfaces can generate spin-dependent scattering as well.

We will now discuss two possible ways to describe the spin-dependent properties inherent to interfaces. These types of modelling can be combined into complete models. Yet other ways to account for more spin-dependent interface properties are, for example, detailed in V. P. Amin and M. D. Stiles, *Phys. Rev.* B **94**, 104419 (2016) and *ibid* 104420 (2016).

■ Spin memory boost

It is possible to introduce the notion of interfacial spin-dependent electronic scattering, and model interface-related changes in the partial DOS. To do so, the

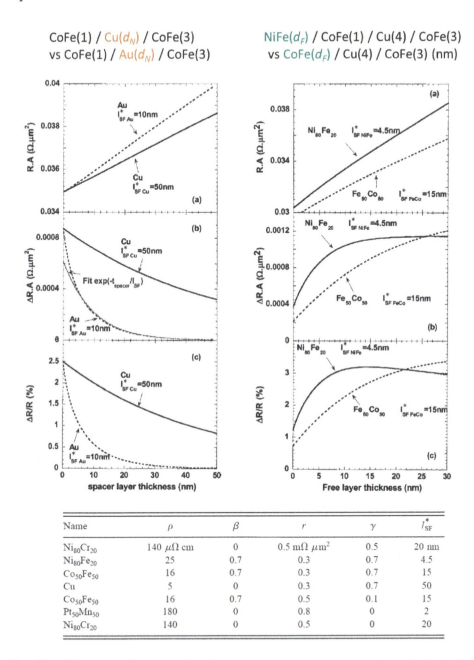

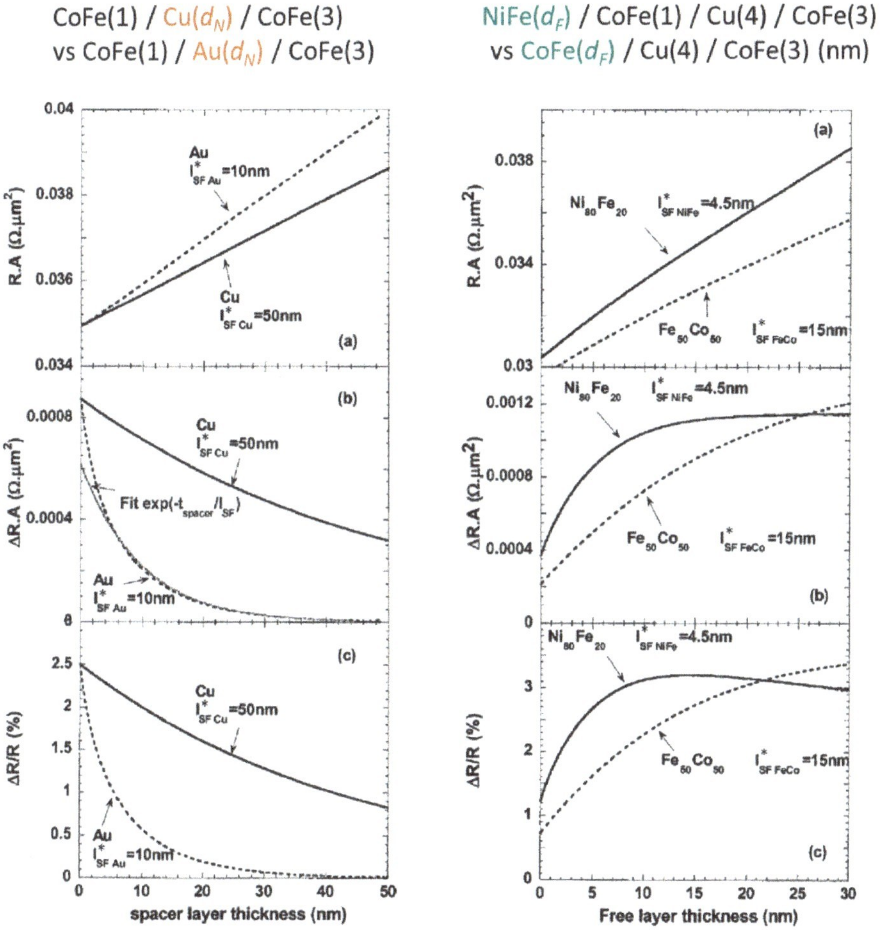

Name	ρ	β	r	γ	l^*_{SF}
$Ni_{80}Cr_{20}$	140 $\mu\Omega$ cm	0	0.5 mΩ μm²	0.5	20 nm
$Ni_{80}Fe_{20}$	25	0.7	0.3	0.7	4.5
$Co_{50}Fe_{50}$	16	0.7	0.3	0.7	15
Cu	5	0	0.3	0.7	50
$Co_{50}Fe_{50}$	16	0.7	0.5	0.1	15
$Pt_{50}Mn_{50}$	180	0	0.8	0	2
$Ni_{80}Cr_{20}$	140	0	0.5	0	20

FIG. 28 – Calculations illustrating how the electron transport parameters in correlation to the layers' thicknesses influence: the resistance area product in the P configuration, the absolute MR area product, and the relative CPP-GMR, in typical F/N/F trilayers. Note: γ is a parameter defined in the next paragraph. It takes into account spin-dependent process at the interface. r is an interfacial resistance area product related to γ. Adapted with permission from AIP Publishing: N. Strelkov *et al.*, *J. App. Phys.* **94**, 3278 (2003). Copyright 2003.

interface is modelled by adding an infinitesimally thin (ferromagnetic) layer displaying a spin asymmetry parameter γ (see figure 29 for numbers). Interface spin-dependent resistance area products are thus introduced in the model as

$$r_{\text{interface}}^{\uparrow(\downarrow)} = 2r_{\text{interface}}^{*}(1-(+)\gamma) \tag{40}$$

with $r_{\text{interface}}^{*} = r_{\text{interface}}/(1-\gamma^2)$.

These notations naturally resemble the ones used for bulk [equation (10) and corresponding text]: $\rho_F^{\uparrow(\downarrow)} = 2\rho_F^*(1-(+)\beta)$. This type of additional interfacial spin-dependent electronic scattering can give rise to an extra polarization of the current. It certainly modifies the spin-coupled interface resistance [equation (37)] and the corresponding voltage drop [T. Valet and A. Fert, *Phys. Rev. B* **48**, 7099 (1993); N. Strelkov *et al.*, *J. App. Phys.* **94**, 3278 (2003)].

Metal pair	$\gamma_{F/N}$	$2AR^*_{F/N}$(fΩm^2)	$\gamma_{F/N}2AR^*_{F/N}$(fΩm^2)
Co/Cu	0.87	1.0	0.9 [4]
Co/Ag	0.85	1.1	0.9 [25]
Fe/Cr	−0.7; −0.59	1.6	1 [95, 96]
Fe/Cu	0.55	1.5	1 [89]
Py/Cu	0.7	1.0	0.7 [71, 79]
Ni/Cu	0.3	0.36	0.1 [90]
Py/Al	0.025	8.5	0.2 [97]
$Co_{90}Fe_{10}$/Al	0.1	10.6	1 [97]
Fe/Al	0.05	8.4	0.4 [97]

Metal pair	Δa/a (%)	$\delta_{F/N}$ or $\delta_{F1/F2}$
Co/Cu	1.8	$0.33^{+0.03}_{-0.08}$ [20]
Co/Ni	0.05	0.35 ± 0.05 [99]
$Co_{90}Fe_{10}$/Cu	1.8	0.19 ± 0.04 [98]
Co/Ru	7	$0.34^{+0.04}_{-0.02}$ [103]

FIG. 29 – Interface spin asymmetry γ and spin-flip δ parameters, which come into play in spintronics effects involving spin-accumulation. Reprinted with permission from Springer Nature: J. Bass, CPP-GMR: Materials and Properties, in Y. Xu *et al.* (eds) *Handbook of Spintronics*, Springer, Dordrecht (2016). Copyright 2016. Note that references from the figure refer to the latter citation and not to references from the present book. See also figure 25.

■ Spin memory loss

An interface can also cause memory loss, or even worse, memory loss.

More seriously, it is also possible to introduce the notion of interfacial spin-dependent spin-flip scattering to account for the fact that only a fraction of the spin current coming from the F layer effectively reaches the N layer. This effect is called spin memory loss (SML). The interface is considered as a thin extra layer with a finite thickness $d_{\text{interface}}$ possessing a finite spin-flip length $l_{sf,\text{interface}}^{*}$ (figure 30). Spin-flipping is modelled by the interface spin-flip parameter $\delta = d_{\text{interface}}/l_{sf,\text{interface}}^{*}$ (see figure 29 for numbers).

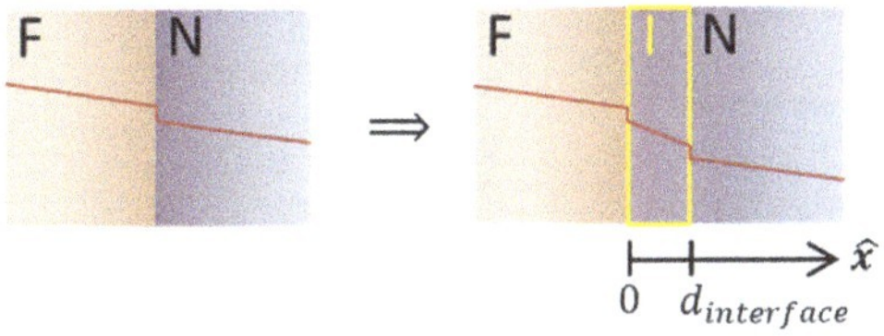

FIG. 30 – Illustrations showing how spin-dependent spin-flip scattering due to the interface (I) itself can be introduced in models, by adding a finite-size thin layer possessing a finite spin-diffusion length $l^*_{sf,\text{interface}}$.

The following ratio is used as a prefactor to weight data obtained from bulk-type calculations

$$R_{\text{SML}} = \frac{J_s(x = d_{\text{interface}})}{J_s(x = 0)} \tag{41}$$

More information on spin memory loss is available for example in J. C. Rojas-Sanchez *et al.*, *Phys. Rev. Lett.* **112**, 106602 (2014).

3.5 Non-Collinearity and Non-Uniformity (Geometric and Magnetic)

We first recall that spin accumulation results from the presence of excess spin density $\delta n^{\uparrow(\downarrow)}$ compared to the steady state condition in isolated layers. It is formulated in the framework of electrochemical potentials by $\mu_s = -\left(\mu^{\uparrow} - \mu^{\downarrow}\right)$, with $\mu^{\uparrow(\downarrow)} = \mu_e + \mu_{n_0} + \frac{\delta n^{\uparrow(\downarrow)}}{N^{\uparrow(\downarrow)}(\varepsilon_F)} = \mu_e + \mu_{n_0} + e^2 D^{\uparrow(\downarrow)} \delta n^{\uparrow(\downarrow)} / \sigma^{\uparrow(\downarrow)}$ (page 36). In uniform systems we have considered so far, δn and subsequently μ_s are scalars. In the most general case, because spin is a vector, $\boldsymbol{\delta n}$ and consequently $\boldsymbol{\mu}_s$ are vectors (exercise 6.8). Hence, spin accumulation is also often described in terms of magnetization imbalance $\boldsymbol{m}_s$, with $\boldsymbol{m}_s = -\mu_B\left[N^{\uparrow}(\varepsilon_F)\boldsymbol{\mu}^{\uparrow} - N^{\downarrow}(\varepsilon_F)\boldsymbol{\mu}^{\downarrow}\right]$. This type of description implicitly recalls that spin accumulation is a vector property, which is crucial when dealing with non-collinearity and non-uniformity. To give an idea of the orders of magnitude, the ratio between spin accumulation and magnetization is $m_s/M \sim 10^{-5} - 10^{-4}$.

An example of non-collinearity is the case where the magnetizations of the F layers in an F/N/F trilayer are neither exactly P nor exactly AP – either by design or due to thermal noise – giving rise to the angular momentum transfer between the flux of angular momentum related to $\boldsymbol{m}_s$ and the layer magnetizations $\boldsymbol{M}$ (chapter 4).

The first example of non-uniformity is the case where the geometry of the system is such that it creates local excess of charge current, at hot spots, and subsequently of spin-current (figure 31). Because the spin accumulation is described by a vector in the most general case, *i.e.*, 3 components of $\boldsymbol{m}_s$, the spin current is described by a 3×3 2nd order tensor, with 3 components of $\boldsymbol{J}_s$ flowing along 3 directions in space,

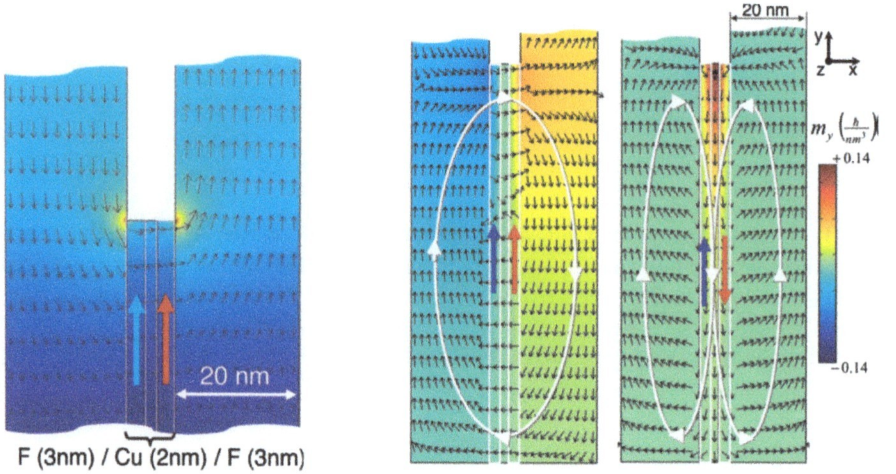

FIG. 31 – Simulations of the transport properties of a magnetoresistive F/N/F nanopillar sandwiched between two extended non-magnetic electrodes. (Left) The color map indicates the amplitude of the charge current J_e (scalar). The arrows indicate the flow of charges $\boldsymbol{J}_e$ (vector). (Right) The color map indicates the amplitude of the y-component of spin accumulation m_y (scalar). The arrows indicate the corresponding flow of y-spins $\boldsymbol{J}_s^y$ (vector), with $\boldsymbol{J}_s^y = J_s^{yx}\hat{\boldsymbol{x}} + J_s^{yy}\hat{\boldsymbol{y}}$. We recall that, an *overcaret* is used for unit vectors, *e.g.* $\hat{\boldsymbol{x}} = \boldsymbol{x}/x$. Reprinted with permission from IEEE: N. Strelkov *et al.*, *IEEE Magn. Lett.* **1**, 3000304 (2010). Copyright 2010.

$$\boldsymbol{J}_s = \overline{\overline{J}}_s = \begin{bmatrix} \boldsymbol{J}_s^{(x)} & \boldsymbol{J}_s^{(y)} & \boldsymbol{J}_s^{(z)} \end{bmatrix} = \begin{bmatrix} J_s^{xx} & J_s^{yx} & J_s^{zx} \\ J_s^{xy} & J_s^{yy} & J_s^{zy} \\ J_s^{xz} & J_s^{yz} & J_s^{zz} \end{bmatrix} \tag{42}$$

where the i and j indices in the $\boldsymbol{J}_s^{(i)}$ vectors and J_s^{ij} scalar coefficients account for the spin and space components, respectively [see also equation (50) for a quantum mechanical treatment].

The second example of non-uniformity concerns the case of magnetic textures like domain walls (DWs), figure 32. The DW magnetoresistance is defined as

$$\frac{\Delta\rho}{\rho} = \frac{\rho_{\text{with DW}} - \rho_{\text{without DW}}}{\rho_{\text{without DW}}} = \frac{\rho_{\text{DW}} - \rho_0}{\rho_0} \tag{43}$$

Spin accumulation and associated resistance can exist in two ways in magnetic textures: outside and in the texture.

■ Outside the magnetic texture

A naive argument would consist of considering the wall as an abrupt transition when the spin-diffusion length l_{sf}^* is much larger than the DW width w_{DW}. Equation (33) obtained for an isolated interface (§3.3) is then used to describe spin accumulation

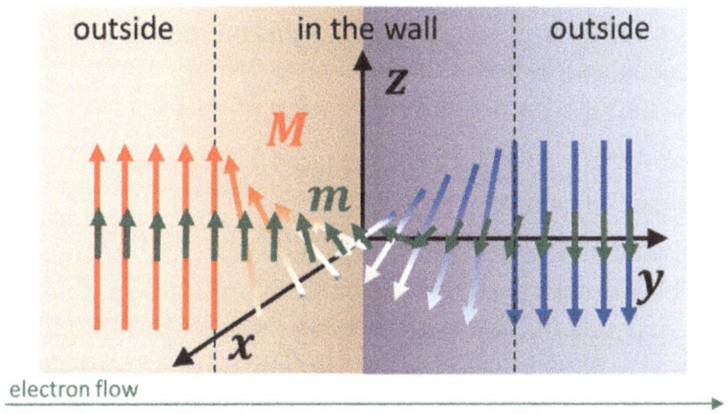

FIG. 32 – Schematics of a Bloch magnetic domain wall illustrating the phenomenon of mistracking between magnetization $\boldsymbol{M}$ and the moment density $\boldsymbol{m}$ of itinerant *s*-electrons, which contributes to domain wall magnetoresistance.

around the wall. The corresponding spin-coupled resistance [equation (37), with $N = F$] serves to estimate the wall resistance. However, this vision is most often wrong. Indeed, looking at it more closely, the wall itself needs to be considered, especially the ability of the moment (spin) density $\boldsymbol{m}$ of the itinerant *s*-electrons to follow magnetization $\boldsymbol{M}$ (figure 32). In the ultimate case when there is perfect tracking, the electron spin sees a homogeneous medium, resulting in no spin accumulation and no wall resistance.

In the most general case, spin-tracking reduces the ability of the wall to generate spin accumulation around it. We note that the term spin-tracking is equivalent to moment-tracking. A full description considers the spin mistracking parameter ξ. In low spin-orbit materials, this parameter can be viewed as the square of the ratio between the exchange length $l_{sd} = \sqrt{D\hbar/J_{sd}}$, and the spin diffusion length l_{sf} (see also §4.2 and §4.3). The change in resistance area product due to spin accumulation around the wall is [E. Šimánek and A. Rebei, *Phys. Rev. B* **71**, 172405 (2005)]

$$\Delta r_{\boldsymbol{m}_s} = \frac{8\sqrt{3}}{9}\beta^2 \rho_F^* \xi^2 \lambda_e \tag{44}$$

This contribution is usually negligible compared to other contributions.

■ In the magnetic texture

A more subtle 'intrinsic' contribution to DW magnetoresistance is directly related to spin mistracking as this gives rise to a local and gradual mixing of eigenstates (figure 32), and therefore opens channels for scattering. This contribution can be of the order of a few percent. The corresponding change in resistance area product is [P. M. Levy and S. Zhang, *Phys. Rev. Lett.* **79**, 5110 (1997)]

$$\Delta r_i = \frac{12}{5}\beta^2 \rho_F^* \xi^2 w_{\mathrm{DW}}\left(1 + \frac{5}{3}\sqrt{1-\beta^2}\right) \tag{45}$$

The case of large spin-orbit materials like semiconductors and the way spin-orbit interactions influence spin mistracking and spin transfer torque in DW (reciprocal effect to DW magnetoresistance) will be mentioned in §4.5.

■ Note about the AMR contribution in magnetic textures

We remark that the spin-accumulation-induced DW magnetoresistances coexist with yet another contribution: the one due to AMR (DWAMR). Its origin is the same as the AMR effect described in §2.6 for a uniform system. It is due to spin-orbit interactions and depends on the angle between magnetization and current. In the case of the Bloch DW shown in figure 32, the DWAMR is zero, $\Delta\rho_{\mathrm{AMR}} = 0$, as the current (along x) is perpendicular to magnetization (in the yz plane), with and without a DW: $\rho_{\mathrm{DW}} = \rho_0$. Exercise 6.2 is devoted to DWAMR.

Summary

In this chapter, we have seen that spin accumulation occurs when electrons flow across an interface, due to the distinct partial densities of current between materials of different types, for example, a ferromagnet ($\beta \neq 0$) and a non-magnet ($\beta = 0$). Spin accumulation is described as the difference between electrochemical potentials $\mu_s = -(\mu^{\uparrow} - \mu^{\downarrow})$. It can be calculated in the frame of the Valet and Fert drift-diffusion model. Relaxation to a steady state and the spatial extension of spin accumulation is regulated by spin-flip scattering, modeled by an *average* spin-diffusion length l_{sf}^*. Spin accumulation gives rise to a spin-coupled interface resistance r_s. A good match between the spin impedance ρl_{sf}^* of the materials on either side of the interface is needed for efficient spin injection. The odd symmetry of spin accumulation with magnetization orientation and stacking order in heterostructures is involved in the GMR effect. Intrinsic contributions due to the interface (spin asymmetry and spin-flip parameters, γ and δ), non-collinearity, and non-uniformity (geometric and magnetic) alter spin accumulation and must be considered for a complete description.

Chapter 4

Transfer of Angular Momentum – STT, Spin Pumping, SOT

In chapter 3, we described how electron transport across an interface creates spin accumulation. In this chapter, we will detail the process by which spin angular momentum (or equivalently magnetic moment) can be transferred from current to magnetization, and the type of spin transfer torque (STT) that this effect generates. The key parameters involved, such as the spin mixing conductance, will be discussed in detail, using a toy quantum mechanical model. How STT terms alter the transport equations, and subsequently the spin accumulation, by providing an additional relaxation channel for the spins (moments) (§3.2) will be tackled. The ability of STT to trigger oscillations and magnetization reversal will be discussed, using the dynamic equations of magnetization. In fact, spin transfer torque STT and giant magnetoresistance GMR (§3.3) are reciprocal effects. In the GMR effect, magnetization alters electron transport due to spin-dependent relaxation. In the STT effect, the electron transport alters magnetization for the same reasons, except that the current necessary to observe the effect is larger, *e.g.*, an order of magnitude larger for STT-induced magnetization reversal than for GMR. Next, guidance on the morphology of STT in non-uniform magnetic textures will be given, echoing §3.5. In a penultimate section, we briefly present how STT is modeled in the frame of the magnetoelectronic circuit theory, which is widely used in current spintronics, and how this theory compares to the toy model presented earlier. A final section introduces spin-orbit torques (SOT), which result from the transfer of angular momentum from the lattice to magnetization, mediated by spin-orbit coupling between the current and the lattice.

4.1 (*sd*-)Coupling, Spin Transfer Torques – STT

Consider a $F_{pinned}/N/F_{free}$ trilayer, in which a current flows perpendicular to the interfaces. The different energy contributions in F_{pinned} are such that the 'pinned' layer is not influenced by the current (the layer may have a large coercive field or should be exchange biased by a third-party antiferromagnetic layer). Conversely, the F_{free}

DOI: 10.1051/978-2-7598-2917-0.c004

magnetization is supposed to be 'free' to oscillate/rotate due to the action of the current.

In line with the description in chapter 3, the s-electrons flowing from the pinned layer to the free layer acquire a spin polarization in the F_{pinned} layer of magnetization $\boldsymbol{M}_{pinned}$ and accumulate at the N/F_{free} interface as shown in figure 33. We describe the incoming spins by a magnetic moment density $\boldsymbol{m}_{pinned} = \boldsymbol{\mathcal{m}}_{pinned}/V$, related to the spin angular momentum density $\boldsymbol{s}_{pinned} = \boldsymbol{\mathcal{s}}_{pinned}/V = -\boldsymbol{m}_{pinned}/|\gamma|$. Naturally, $\boldsymbol{m}_{pinned}$ relates to the current polarization in F_{pinned} and to spin accumulation at the N/F_{free} interface.

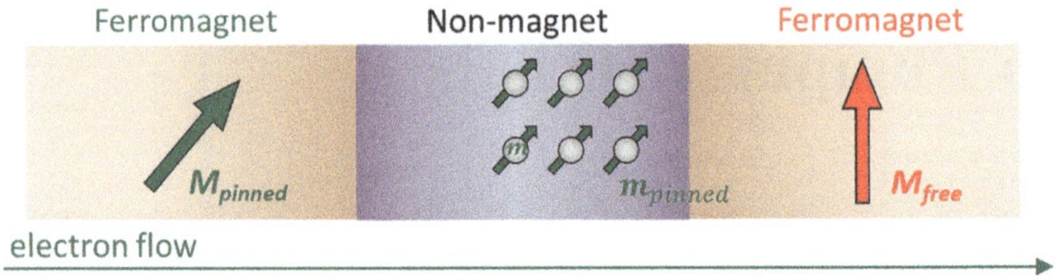

FIG. 33 – Illustration of a pinned-ferromagnet/non-magnet/free-ferromagnet trilayer, when the magnetizations $\boldsymbol{M}_{pinned}$ and $\boldsymbol{M}_{free}$ of the ferromagnets are misaligned. The moments incoming on the free-ferromagnet are polarized in the pinned-ferromagnet. They are described by their magnetic moment density $\boldsymbol{m}_{pinned}$ related to the spin angular momentum density $\boldsymbol{s}_{pinned}$ as $\boldsymbol{m}_{pinned} = -|\gamma|\boldsymbol{s}_{pinned}$.

The incoming $\boldsymbol{m}_{pinned}$ (related to s-electrons) and the magnetization $\boldsymbol{M}_{free}$ (related to d-electrons) are coupled *via* the *sd* exchange interactions, which is expressed by the following Hamiltonian: $\mathcal{H}_{sd} = -J_{sd}\widehat{\boldsymbol{m}}_{pinned}.\widehat{\boldsymbol{M}}_{free}$, with the exchange constant J_{sd}. Consequently, in the reference frame of the layer magnetization, $\boldsymbol{M}_{free}$ experiences an effective field $\boldsymbol{H}_{\boldsymbol{m}_{pinned}\rightarrow \boldsymbol{M}_{free}} \propto J_{sd}\boldsymbol{m}_{pinned}$ and a torque $\boldsymbol{T}_{\boldsymbol{m}_{pinned}\rightarrow \boldsymbol{M}_{free}} \propto -\frac{1}{|\gamma|}\frac{d\boldsymbol{M}_{free}}{dt}$, called the spin transfer torque, STT. In a more detailed description, we will see that, in fact, what matters for spin transfer torque is the <u>flux</u> of $\boldsymbol{m}_{pinned}$, or, equivalently, up to a sign, the flux of spin angular momentum $\boldsymbol{s}_{pinned}$. Note that in the following we give the expression of $-\boldsymbol{T}$ because $\frac{d\boldsymbol{M}}{dt}$ goes like $-|\gamma|\boldsymbol{T}$ (see also equation 65).

Only projections of $\boldsymbol{m}_{pinned}$ perpendicular to $\boldsymbol{M}_{free}$ give rise to a torque. Therefore, the STT is often written in terms of the following two orthogonal components: the damping-like (DL) torque $-\boldsymbol{T}_{DL} = \frac{\tau_{DL}}{M_S^2}\boldsymbol{M} \times (\boldsymbol{m} \times \boldsymbol{M})$ and the field-like (FL) torque $-\boldsymbol{T}_{FL} = \frac{\tau_{FL}}{M_S}\boldsymbol{m} \times \boldsymbol{M}$, in J.m^{-3}. The terms τ_{DL} and τ_{FL} are the DL and FL magnetic flux density (or magnetic B-field) amplitudes, in tesla, for direct comparison with the action of a flux density $\mu_0\boldsymbol{H}$ oriented along $(\boldsymbol{M} \times \boldsymbol{m})$ and $\boldsymbol{m}$, respectively [figure 34, also equation (46)]. We took $\boldsymbol{M} = \boldsymbol{M}_{free}$ and $\boldsymbol{m} = \boldsymbol{m}_{pinned}$ for simplifications. Actually, the expressions in terms of the angular momentum density $\boldsymbol{s} = -\frac{1}{|\gamma|}\boldsymbol{m}$ of the incoming current are also encountered in the literature: $-\boldsymbol{T}_{DL} = |\gamma|\frac{\tau_{DL}}{M_S^2}\boldsymbol{M} \times (\boldsymbol{M} \times \boldsymbol{s})$ and $-\boldsymbol{T}_{FL} = |\gamma|\frac{\tau_{FL}}{M_S}\boldsymbol{M} \times \boldsymbol{s}$. Most often, these expressions are used

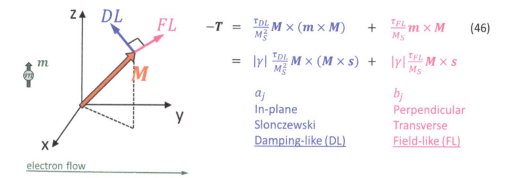

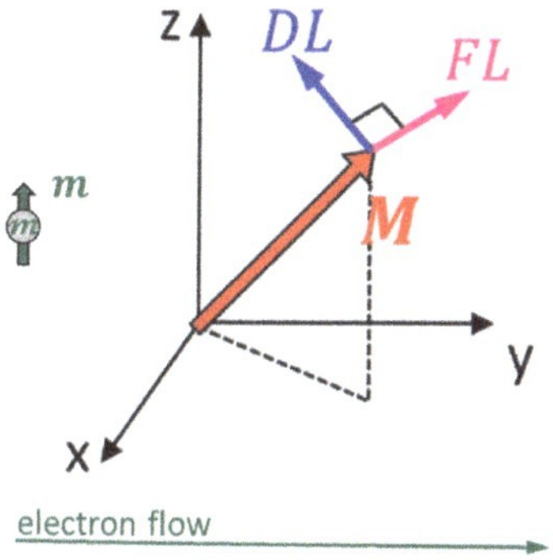

$$-\boldsymbol{T} = \frac{\tau_{DL}}{M_S^2}\boldsymbol{M}\times(\boldsymbol{m}\times\boldsymbol{M}) + \frac{\tau_{FL}}{M_S}\boldsymbol{m}\times\boldsymbol{M} = |\gamma|\frac{\tau_{DL}}{M_S^2}\boldsymbol{M}\times(\boldsymbol{M}\times\boldsymbol{s}) + |\gamma|\frac{\tau_{FL}}{M_S}\boldsymbol{M}\times\boldsymbol{s} \quad (46)$$

a_j	b_j
In-plane	Perpendicular
Slonczewski	Transverse
Damping-like (DL)	Field-like (FL)

FIG. 34 – Representation of the damping-like (DL) and field-like (FL) contributions to $d\boldsymbol{M}/dt = -|\gamma|\boldsymbol{T}$ due to torques acting on a magnetization $\boldsymbol{M}$, originating from transfer of angular momentum with incoming electrons of moment density $\boldsymbol{m}$, and spin density $\boldsymbol{s} = -\boldsymbol{m}/|\gamma|$. The DL torque orients $\boldsymbol{M}$ towards $\boldsymbol{m}$, or equivalently away from $\boldsymbol{s}$, and the FL torque induces a precession anticlockwise viewed from above.

with normalized quantities. In this book, we used the non-normalized values to make it clear in terms of units and conversion from one notion to another.

The above type of expressions for the STT are the most general ones, as they consider the universal case of polarized electrons incoming on a magnetization $\boldsymbol{M}$, be the polarization of the electrons expressed as $\boldsymbol{s}$ or $\boldsymbol{m}$, and whatever the effect at its origin.

In fact, the direction of $\boldsymbol{s}$ or $\boldsymbol{m}$ depends on the effect that gave rise to spin polarization. Therefore, the directions of FL and DL torques can be reversed depending on the initial direction of $\boldsymbol{m}$. For STT, $\boldsymbol{m} \parallel \boldsymbol{M}_{\text{pinned}}$, but in some other structures for spintronics, effects arising from spin-orbit coupling, like the Rashba effect, can give rise to a different orientation of the spin polarization and, subsequently, to a different orientation of the torques. These torques are called spin-orbit torques (SOT) and will be introduced in §4.7.

In articles on STT, in GMR structures, because the polarization of the current $\boldsymbol{m}_{\text{pinned}}$ relates to $\boldsymbol{M}_{\text{pinned}}$, it is common to find the torques expressed in terms of $\boldsymbol{M}_{\text{pinned}}$: $-\boldsymbol{T}_{\text{DL}} = \frac{\tau_{\text{DL}}}{M_S^2}\boldsymbol{M}_{\text{free}} \times (\boldsymbol{M}_{\text{pinned}} \times \boldsymbol{M}_{\text{free}})$ and $-\boldsymbol{T}_{\text{FL}} = \frac{\tau_{\text{FL}}}{M_S}\boldsymbol{M}_{\text{pinned}} \times \boldsymbol{M}_{\text{free}}$.

When more than one terminology can be found, the names damping-like (DL) and field-like (FL) are now accepted. The origin of such a name is direct from the magnetization dynamics framework [equation (65)], which will be detailed in §4.4. The sign convention will be detailed in §4.2.

We note that, the total magnitude of the torque terms depends on materials properties (*e.g.*, magnetic volume $M_S d_F$, and polarization β), the geometry, and the current amplitude and direction (J_e) [equation (65)]. About the orientation of the torques with regards to the current direction, the rule is as follows: when an electron of magnetic moment density $\boldsymbol{m}$ (and spin density $\boldsymbol{s}$ pointing in the opposite direction) flows toward a magnetization $\boldsymbol{M}$, the DL torque orients $\boldsymbol{M}$ towards $\boldsymbol{m}$, or equivalently away from $\boldsymbol{s}$, and the FL torque induces a precession anticlockwise viewed from above (figure 34).

Finally, we remark that, in the literature, $\boldsymbol{m}$ is used either to define a property, in $\mathrm{A.m^{-1}}$, units of magnetization, as throughout this textbook, or for the dimensionless magnetization unit vector, in which case $\boldsymbol{m} = \frac{\boldsymbol{M}}{M_S}$, resulting in some possible misunderstanding and a factor M_S difference is some of the equations depending on the articles and books consulted. In this textbook, we use an *overcaret* for unit vectors, *i.e.*, $\widehat{\boldsymbol{M}} = \frac{\boldsymbol{M}}{M_S}$ (see symbols and units page 150).

The purpose of what follows is to determine, by use of a toy quantum mechanical model:

- **what governs the transfer of angular momentum by STT**, and
- **how magnetization is influenced by the torque *T*.**

4.2 Toy Quantum Mechanical Model for STT, Angular Momentum Flow

■ Reminder

Before presenting the quantum mechanical model, it is important to recall the quantum mechanical treatment of charge and spin, which we will use.

Consider a spinor ψ for the full description of the spin-electron state, *i.e.*, a spin-half electron wave function.

The observable density of charge ρ is described by the product between the electron charge $-e$ and the density probability, and the observable density of charge current $\boldsymbol{J}_e$ additionally involves the velocity operator $\mathbf{v}$

$$\rho = -e\langle\psi|\psi\rangle \tag{47}$$

$$\boldsymbol{J}_e = -e\mathcal{R}e(\langle\psi|\mathbf{v}|\psi\rangle) = \frac{e\hbar}{m_e}\Im m(\langle\psi|\boldsymbol{\nabla}|\psi\rangle) \tag{48}$$

Regarding the spin, the density of spin angular momentum is described by

$$\mathbf{s} = \langle\psi|\mathbf{S}|\psi\rangle = \frac{\hbar}{2}\langle\psi|\boldsymbol{\sigma}|\psi\rangle \tag{49}$$

and the density of spin current by

$$\boldsymbol{Q} = \frac{\hbar}{2e}\boldsymbol{J}_s = \mathcal{R}e(\langle\psi|\mathbf{S}\otimes\mathbf{v}|\psi\rangle) = -\frac{\hbar^2}{2m_e}\Im m(\langle\psi|\boldsymbol{\sigma}\otimes\boldsymbol{\nabla}|\psi\rangle) \tag{50}$$

Both expressions involve the use of the spin operator $\mathbf{S}$. It is related to the Pauli operator $\boldsymbol{\sigma}$ by $\mathbf{S} = \frac{\hbar}{2}\boldsymbol{\sigma}$, with $\sigma^x = \begin{pmatrix} 0 & 1 \\ 1 & 0 \end{pmatrix}$, $\sigma^y = \begin{pmatrix} 0 & -i \\ i & 0 \end{pmatrix}$, and $\sigma^z = \begin{pmatrix} 1 & 0 \\ 0 & -1 \end{pmatrix}$; z being the quantization axis. The expression for the spin current density involves a tensor product $\otimes$ between the spin operator $\mathbf{S}$ from the spin Hilbert space, and the velocity operator **v** from the coordinate Hilbert space (see also §3.5 for the classical definition of the spin current tensor).

■ Toy quantum mechanical model

We now consider an incoming spin density at the N/F_{free} interface, resulting from an initial spin polarization in the F_{pinned} layer of a $F_{pinned}/N/F_{free}$ trilayer (figure 33). The magnetization of the F_{pinned} layer is angled by θ, whereas that of the F_{free} layer is set along the z axis (making it the quantization axis). For the full description, please refer to M. D. Stiles and J. Miltat, *Top. Appl. Phys.* **101**, 225 (2006).

Dynamic spin transport is represented by an incoming electron-spin with a state $|\psi_i\rangle$ – carrying a density of angular momentum $\boldsymbol{s}$ (figure 35). $|\psi_i\rangle$ is a linear combination of the $|\Uparrow\rangle$ and $|\Downarrow\rangle$ eigenstates: $|\psi_i\rangle = \frac{1}{\sqrt{V}}\left(\cos\frac{\theta}{2}|\Uparrow\rangle + \sin\frac{\theta}{2}|\Downarrow\rangle\right)$. We recall that the $\uparrow$ and $\downarrow$ symbols refer exclusively to the orientation of $\boldsymbol{m}$, and the $\Uparrow$ and $\Downarrow$ symbols refer exclusively to the orientation of $\boldsymbol{s}$.

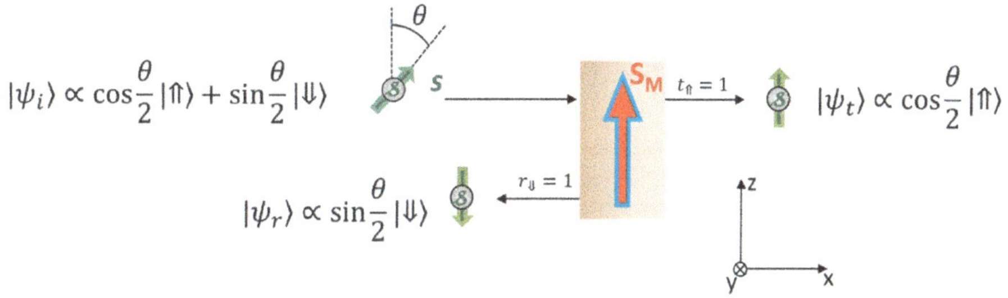

FIG. 35 – Schematics corresponding to a toy model for STT (see text), illustrating how an incoming electron with wavefunction $|\psi_i\rangle$ – carrying angular momentum $\boldsymbol{s}$ – transfers angular momentum with magnetization – having angular momentum density $\boldsymbol{S_M}$. $|\psi_t\rangle$ and $|\psi_r\rangle$ are the transmitted and reflected wavefunctions. $t_\Uparrow$ and $r_\Downarrow$ are the transmission and reflection coefficients for the $|\Uparrow\rangle$ and $|\Downarrow\rangle$ eigenstates, respectively. Normalization to volume $1/\sqrt{V}$ and propagation terms e^{ikx} and e^{-ikx} were omitted to simplify the scheme.

Upon impinging on the F_{free} layer, represented by the macrospin angular momentum $V\boldsymbol{S_M} = -\frac{1}{|\gamma|}V\boldsymbol{M}$, with V the volume, the incoming wavefunction is filtered for the $|\Uparrow\rangle$ and $|\Downarrow\rangle$ eigenstates (§2.4). Filtering is expressed in terms of transmission and reflection coefficients for each state, $t_{\Uparrow(\Downarrow)}$ and $r_{\Uparrow(\Downarrow)}$. These parameters take into account the *sd* exchange interactions between $\boldsymbol{s}$ and $\boldsymbol{S_M}$, which are at the origin of the transfer of angular momentum. In this toy model, we consider that the $|\Uparrow\rangle$ eigenstate is transmitted, $t_{\Uparrow(\Downarrow)} = 1(0)$ and the $|\Downarrow\rangle$ eigenstate is reflected, $r_{\Uparrow(\Downarrow)} = 0(1)$. The resulting transmitted and reflected states are thus expressed as follows: $|\psi_t\rangle = \frac{1}{\sqrt{V}}\cos\frac{\theta}{2}|\Uparrow\rangle$ and $|\psi_r\rangle = \frac{1}{\sqrt{V}}\sin\frac{\theta}{2}|\Downarrow\rangle$.

Note that, the use of $\frac{\theta}{2}$ and not of θ comes from the fact that the eigenstates of opposite (π) spins are represented on an orthogonal $\left(\frac{\pi}{2}\right)$ basis. $1/\sqrt{V}$ is used to normalize the volume.

Combining equation (49) with the incoming $|\psi_i\rangle$, transmitted $|\psi_t\rangle$, and reflected $|\psi_r\rangle$ states, it is possible to express[1] the incoming $s_i^{x(z)}$, transmitted $s_t^{x(z)}$, and reflected $s_r^{x(z)}$ densities of spins. They are decomposed along the z quantization axis and its orthogonal direction x by use of $s_{i(t)(r)}^{x(z)} = \langle\psi_{i(t)(r)}|S^{x(z)}|\psi_{i(t)(r)}\rangle$.

One obtains $s_i^x = \frac{\hbar}{2V}\sin\theta$ and $s_i^z = \frac{\hbar}{2V}\cos\theta$ for the incoming density of spin. The final density of spin after interaction through transfer of angular momentum between the incoming electron of density of angular momentum $\boldsymbol{s}$ and the ferromagnet macrospin $\boldsymbol{S_M}$, is the sum of the transmitted and reflected densities of spins. One obtains $s_f^x = s_t^x + s_r^x = 0 + 0 = 0$ and $s_f^z = s_t^z + s_r^z = \frac{\hbar}{2V}\cos^2\frac{\theta}{2} - \frac{\hbar}{2V}\sin^2\frac{\theta}{2} = \frac{\hbar}{2V}\cos\theta$.

The loss of angular momentum for the incoming electron can be expressed as $\Delta s^x = \left(s_t^x + s_r^x\right) - s_i^x = -\frac{\hbar}{2V}\sin\theta$ and $\Delta s^z = \left(s_t^z + s_r^z\right) - s_i^z = 0$. Using conservation of total angular momentum, we obtain $\Delta s + \Delta S_M = 0$. The correlated gain of angular momentum for the magnetic layer is thus expressed as $\Delta S_M = \Delta S_M^x = \frac{\hbar}{2V}\sin\theta$.

This angular momentum transfer from the spin of incoming electrons to magnetization can be rewritten as

$$\boldsymbol{\Delta S_M} = \Delta S_M^x \widehat{\boldsymbol{x}} = \frac{\hbar}{2V}\sin\theta\widehat{\boldsymbol{x}} = \boldsymbol{s}_\perp \tag{51}$$

It is said that the transverse component of spin angular momentum is absorbed. This eventually alters magnetization (figure 36), as the result of a torque $\boldsymbol{T}_V$, acting

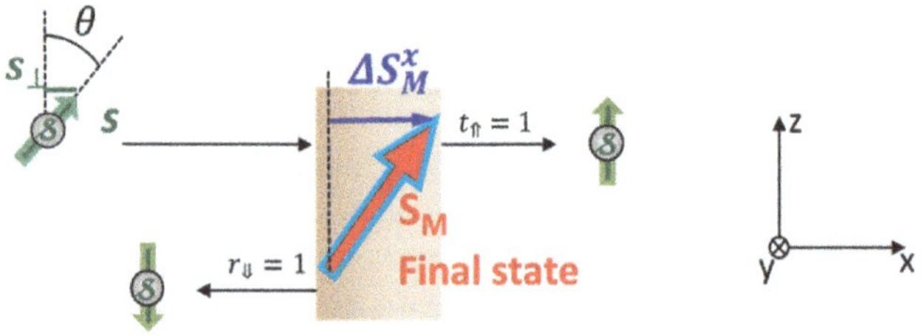

FIG. 36 – Follow-up schematics corresponding to figure 35 after transfer of angular momentum. The transverse component of the spin angular momentum density $\boldsymbol{s}_\perp$ has been absorbed by the macrospin $\boldsymbol{S_M}$, which has thus gained $\Delta S_M^x \widehat{\boldsymbol{x}} = \boldsymbol{s}_\perp$. In this example, the torque is damping-like.

[1] In this footnote, we detail the calculation of the incoming density of spin along x, as an example. Calculation of the other incoming s_i^z, transmitted $s_t^{x(z)}$ and reflected $s_r^{x(z)}$ densities of spins is then straightforward. The Bra and Ket state vectors are represented by column vectors and their conjugate transpose, respectively: $\langle\Uparrow| = (1 \quad 0)$, $\langle\Downarrow| = (0 \quad 1)$, $|\Uparrow\rangle = \begin{pmatrix}1\\0\end{pmatrix}$, and $|\Downarrow\rangle = \begin{pmatrix}0\\1\end{pmatrix}$, as it is the use in quantum mechanics. Therefore, $s_i^x = \langle\psi_i|S^x|\psi_i\rangle = \left(\cos\frac{\theta}{2} \quad \sin\frac{\theta}{2}\right)\frac{\hbar}{2}\begin{pmatrix}0 & 1\\1 & 0\end{pmatrix}\begin{pmatrix}\cos\frac{\theta}{2}\\ \sin\frac{\theta}{2}\end{pmatrix} = \frac{\hbar}{2}2\cos\frac{\theta}{2}\sin\frac{\theta}{2} = \frac{\hbar}{2}\sin\theta$. We note that normalization to volume and the propagation terms were omitted to facilitate calculations. Those terms can be included afterwards.

on the moment $V\boldsymbol{S_M}$. In turn, the spin transfer torque $\boldsymbol{T}_V$ can be directly calculated[2] from the spin current flow (here along x) entering (incoming $\boldsymbol{Q}_i$) and leaving (reflected $\boldsymbol{Q}_r$ and transmitted $\boldsymbol{Q}_t$) the volume V, and acting on an area A. It is

$$\boldsymbol{T}_V = -\iint \widehat{\boldsymbol{x}} \cdot \boldsymbol{Q} dA = A\widehat{\boldsymbol{x}} \cdot (\boldsymbol{Q}_i + \boldsymbol{Q}_r - \boldsymbol{Q}_t) \tag{52}$$

$$\boldsymbol{T}_V = \frac{A}{V}\frac{\hbar^2 k}{2m_e}\sin\theta\widehat{\boldsymbol{x}} = \frac{A}{V}\frac{\hbar^2 k}{2m_e}\widehat{\boldsymbol{S}}_M \times \left(\widehat{\boldsymbol{s}} \times \widehat{\boldsymbol{S}}_M\right) \tag{53}$$

Comparing equations (53) and (46) we see that $\boldsymbol{T}_V$ is damping-like only.

■ Refined quantum mechanical model

In refined models, other ingredients that contribute to the process of transfer of angular momentum are introduced (figure 37). We will now describe these ingredients and how they contribute to both a damping-like and a field-like STT.

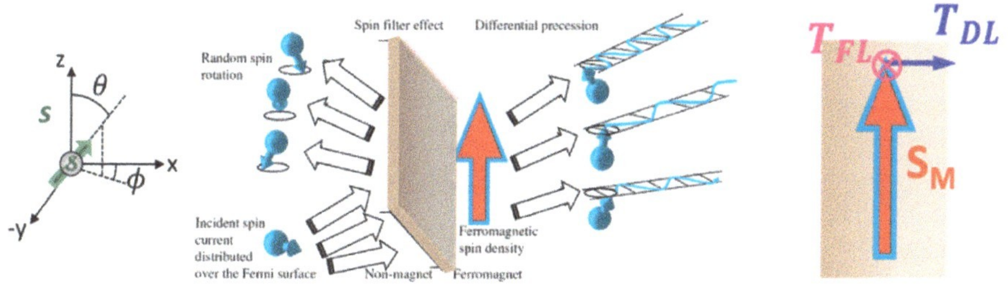

FIG. 37 – Illustration of some of the ingredients involved and introduced in models in order to describe STT more accurately. Those ingredients are: spin filtering due to spin-dependent scattering (§2.4); spin rotation and dephasing upon reflection; and spin precession upon transmission in the ferromagnet. When the ingredients are introduced, both damping-like ($\boldsymbol{T}_{\mathrm{DL}}$) and field-like ($\boldsymbol{T}_{\mathrm{FL}}$) terms contribute to the torque. Reprinted with permission from Springer Nature: M. D. Stiles and J. Miltat, *Top. Appl. Phys.* **101**, 225 (2006). Copyright 2006.

Some of the ingredients introduced to explain the physical origin of STT more accurately are (figure 37):

---

[2]The density of spin current is given by $\boldsymbol{Q} = -\frac{\hbar}{m_e}\Im m(\langle\psi|\boldsymbol{S}\cdot\boldsymbol{\nabla}|\psi\rangle)$ with $Q^{ab} = -\frac{\hbar}{m_e}\Im m(\langle\psi|\boldsymbol{S}^a\cdot\boldsymbol{\nabla}_b|\psi\rangle) = -\frac{\hbar k_b}{m_e}\langle\psi|\boldsymbol{S}^a|\psi\rangle = -\frac{\hbar k_b}{m_e}s^a$, where the element Q^{ab} represents the flow along the direction b of spins polarized along a. Based on the geometry of the problem, we have $\boldsymbol{Q} = \overline{\overline{Q}} = -\frac{\hbar k_x}{m_e}\begin{bmatrix} s^x & s^y & s^z \\ 0 & 0 & 0 \\ 0 & 0 & 0 \end{bmatrix}$ and $\widehat{\boldsymbol{x}}\cdot\boldsymbol{Q} = \frac{\hbar k_x}{m_e}(s^x + s^y + s^z)\widehat{\boldsymbol{x}}$. The torque is therefore given by $\boldsymbol{T}_V = A\widehat{\boldsymbol{x}}\cdot(\boldsymbol{Q}_i + \boldsymbol{Q}_r - \boldsymbol{Q}_t) = -\frac{\hbar k_x}{m_e}(\Delta s^x + \Delta s^y + \Delta s^z)\widehat{\boldsymbol{x}} = \frac{A\hbar^2 k_x}{V 2m_e}\sin\theta\widehat{\boldsymbol{x}}$.

- **spin filtering** (§2.4): spin-dependent transmission ($t_{\Uparrow(\Downarrow)}$) and reflection ($r_{\Uparrow(\Downarrow)}$) coefficients. Initially, these are real numbers that are not necessarily set to 0 or 1 like in the toy model discussed above. They satisfy $t^2_{\Uparrow(\Downarrow)} + r^2_{\Uparrow(\Downarrow)} = 1$.
- **spin rotation (ϕ) and dephasing ($\delta\phi$) upon reflection:** $e^{i(\phi+\delta\phi)}$, leading to a mixing of the eigenstates and represented nowadays by a complex number parameter known as the spin mixing conductance ($G^{\uparrow\downarrow}$). The terminology mixing is naturally the same as that used in §2.6, but the origin of the mixing differs. The term spin mixing refers interchangeably to spin and moment mixing ($G^{\Uparrow\Downarrow} = G^{\uparrow\downarrow}$).
- **spin precession upon transmission in the ferromagnet:** $e^{i\left(k_{\Uparrow}-k_{\Downarrow}\right)x}$, influencing torques and spin accumulation.

The ingredients mentioned above are taken into account in models, in a combined manner, by considering complex number transmission $t_{\Uparrow(\Downarrow)}$ and reflection $r_{\Uparrow(\Downarrow)}$ coefficients. This now enables the spin to exit the (x, z) plane. As an example, the calculated density of reflected spin current in this refined model [M. D. Stiles and J. Miltat, *Top. Appl. Phys.* **101**, 225 (2006)] reads

$$Q_r^{xx} = -\frac{\hbar k_x}{m_e}\sin\theta \mathcal{R}e\left(r^*_{\Uparrow} r_{\Downarrow} e^{i\phi}\right) \tag{54}$$

$$Q_r^{yx} = -\frac{\hbar k_x}{m_e}\sin\theta \Im m\left(r^*_{\Uparrow} r_{\Downarrow} e^{i\phi}\right) \tag{55}$$

where Q^{ab} represents the flow along the direction b of spins polarized along a, and with, by definition, $r^*_{\Uparrow} r_{\Downarrow} = \left|r^*_{\Uparrow} r_{\Downarrow}\right| e^{i\delta\phi}$.

- **Spin mixing conductance**

This latter term $r^*_{\Uparrow} r_{\Downarrow}$ describes spin dephasing upon reflection, due to reflection along axes transverse to $\boldsymbol{S_M} \parallel \hat{z}$. Actually, $r_\perp = \mathcal{R}e\left(r_{\Uparrow} r^*_{\Downarrow}\right)$ and $r_\times = \Im m\left(r_{\Uparrow} r^*_{\Downarrow}\right)$ describe reflection along the two axes transverse to $\boldsymbol{S_M}$. They are called the transverse reflection coefficients. We will see in a few lines that this complex number is involved in the definition of the parameter called the spin mixing conductance $G^{\uparrow\downarrow} \sim \frac{e^2}{h}\left(1 - r^*_{\Uparrow} r_{\Downarrow}\right)$ [equation (80)], which is encountered whenever there is an *sd*-type transfer of angular momentum. The real $G_r^{\uparrow\downarrow} = \mathcal{R}e\left(G^{\uparrow\downarrow}\right)$ and imaginary $G_i^{\uparrow\downarrow} = \Im m\left(G^{\uparrow\downarrow}\right)$ parts of the spin mixing conductance describe the spin current that is polarized along the two axes transverse to $\boldsymbol{S_M}$ and that is absorbed by the ferromagnet, therefore creating torques. We recall that no spin-flip process is involved, and that the terminology 'mixing' accounts for a mixing of the eigenstates. We also note that the spin mixing conductance is most often used per unit area A per quantum conductance per spin channel $G_0/2$: $g^{\uparrow\downarrow} = \frac{G^{\uparrow\downarrow}}{AG_0/2}$. Typical values are a few tenth of nm^{-2}.

Complete calculation [M. D. Stiles and J. Miltat, *Top. Appl. Phys.* **101**, 225 (2006)] shows that, when dephasing and precession are introduced by considering complex transmission and reflection coefficients, both DL and FL terms contribute to the torque

$$\boldsymbol{T}_V = \frac{A}{V}\frac{\hbar^2 k}{2m_e}\sin\theta\Big[\Big(1 - \mathcal{R}e\Big(t_{\Uparrow}t_{\Downarrow}^* + r_{\Uparrow}r_{\Downarrow}^*\Big)\Big)\widehat{\boldsymbol{x}} - \Im m\Big(t_{\Uparrow}t_{\Downarrow}^* + r_{\Uparrow}r_{\Downarrow}^*\Big)\widehat{\boldsymbol{y}}\Big] \tag{56}$$

$$\boldsymbol{T}_V = \frac{A}{V}\frac{\hbar^2 k}{2m_e}\Big[g_r^{\uparrow\downarrow}\widehat{\boldsymbol{S}}_M \times \Big(\widehat{\boldsymbol{s}} \times \widehat{\boldsymbol{S}}_M\Big) + g_i^{\uparrow\downarrow}\widehat{\boldsymbol{S}}_M \times \widehat{\boldsymbol{s}}\Big] \tag{57}$$

with $g^{\uparrow\downarrow} = 1 - \Big(t_{\Uparrow}t_{\Downarrow}^* + r_{\Uparrow}r_{\Downarrow}^*\Big)$ the spin mixing coefficient.

From equation (56), we note that $T_V = 0$ if $t_{\Uparrow} = t_{\Downarrow}$ and $r_{\Uparrow} = r_{\Downarrow}$. Indeed, in this case, there is no spin filtering. We also note that, $T_V = 0$ if $\theta = 0$ or $\theta = \pi$, when there is collinearity between the polarizing layer F_{pinned} and the free layer F_{free} (figure 33) of the $F_{pinned}/N/F_{free}$ trilayer. In practice, thermal noise always alters collinearity.

The torque is maximum for $\theta = \pi/2$. However, equation (56) considers the torque created by one type of electron wave only. The actual torque created by all electrons can be obtained by summation over the Fermi surface of the N layer, corresponding to all possible incident wave vectors. With this complexity, reflected (dephased) spins interact with incoming spins and this interaction affects the initial spin polarization. This results in a non-trivial θ-dependence of STT, and especially shifts the maximum torque to an angle larger than $\theta = \pi/2$ (figure 38). In addition,

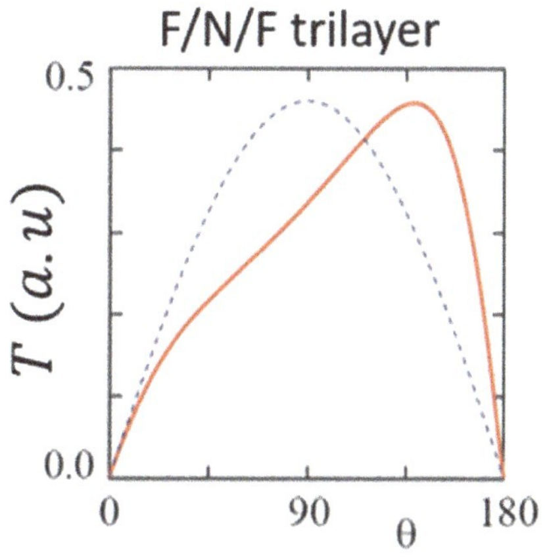

FIG. 38 – Angular(θ)-dependence of the torque T. Spin rotation and dephasing upon reflection affect spin polarization, contributing to non-trivial θ-dependence of T (full line). This trend deviates from the sinθ-dependence (dashed line) anticipated from a simplified model [equation (56)]. Adapted with permission from Springer Nature: M. D. Stiles and J. Miltat, *Top. Appl. Phys.* **101**, 225 (2006). Copyright 2006.

with this complexity, it is possible to explain experimental data showing that, in metallic structures, the FL term is small, less than 5% of the DL term. In this case, some transverse components average out: $\mathcal{Re}\left(t_{\Uparrow}t_{\Downarrow}^{*}\right) = \mathfrak{Im}\left(t_{\Uparrow}t_{\Downarrow}^{*}\right) \approx 0$ and $\mathfrak{Im}\left(r_{\Uparrow}r_{\Downarrow}^{*}\right) \ll \mathcal{Re}\left(t_{\Uparrow}t_{\Downarrow}^{*} + r_{\Uparrow}r_{\Downarrow}^{*}\right)$. This contrasts with tunnel junctions, where the k-selection upon tunneling reduces the effect of averaging. In this latter case, the FL term is about 30% of the DL term.

■ Sign convention for the torques

From equation (57), we get

$$V\frac{d\boldsymbol{S}_M}{dt} = \boldsymbol{T}_V \tag{58}$$

$$\frac{d\boldsymbol{S}_M}{dt} = \frac{A}{V^2}\frac{\hbar^2 k}{2m_e}\left[g_r^{\uparrow\downarrow}\widehat{\boldsymbol{S}}_M \times \left(\widehat{\boldsymbol{s}} \times \widehat{\boldsymbol{S}}_M\right) + g_i^{\uparrow\downarrow}\widehat{\boldsymbol{S}}_M \times \widehat{\boldsymbol{s}}\right] \tag{59}$$

The torque $\boldsymbol{T}_V$ is in N.m ≡ J, because it is the torque of a spin angular momentum density $\boldsymbol{s}$ acting on a macrospin $V\boldsymbol{S}_M$. The actual torque $\boldsymbol{T}$ on $\boldsymbol{S}_M \propto \boldsymbol{M}$ requires dividing by V, in which case the torque is indeed expressed in J.m^{-3}. We also note that the torque is effective over a finite length governed by sd-exchange and spin-orbit coupling (§3.5, §4.3, and §4.5). Therefore, the volume V is in fact an effective volume and not necessarily the total volume of the ferromagnetic free layer. Sometimes, the notation V_{eff} is preferred.

The sign of $\boldsymbol{T}$ with respect to magnetization is obtained from equation (59) with $\boldsymbol{S}_M = -\frac{1}{|\gamma|}\boldsymbol{M}$, *i.e.*, $\widehat{\boldsymbol{S}}_M = -\widehat{\boldsymbol{M}}$. One gets

$$\frac{d\boldsymbol{M}}{dt} = |\gamma|\frac{A}{V^2}\frac{\hbar^2 k}{2m_e}\left[g_r^{\uparrow\downarrow}\widehat{\boldsymbol{M}} \times \left(\widehat{\boldsymbol{M}} \times \widehat{\boldsymbol{s}}\right) + g_i^{\uparrow\downarrow}\widehat{\boldsymbol{M}} \times \widehat{\boldsymbol{s}}\right] \tag{60}$$

Equation (60) thus takes the form

$$\frac{d\boldsymbol{M}}{dt} = -|\gamma|\boldsymbol{T}_{\text{DL}} - |\gamma|\boldsymbol{T}_{\text{FL}} \tag{61}$$

with

$$-\boldsymbol{T}_{\text{DL}} = |\gamma|\frac{\tau_{\text{DL}}}{M_S^2}\boldsymbol{M} \times (\boldsymbol{M} \times \boldsymbol{s}) = \frac{\tau_{\text{DL}}}{M_S^2}\boldsymbol{M} \times (\boldsymbol{m} \times \boldsymbol{M}) \tag{62}$$

$$-\boldsymbol{T}_{\text{FL}} = |\gamma|\frac{\tau_{\text{FL}}}{M_S}(\boldsymbol{M} \times \boldsymbol{s}) = \frac{\tau_{\text{FL}}}{M_S}(\boldsymbol{m} \times \boldsymbol{M}) \tag{63}$$

where τ_{DL} and τ_{FL} are the DL and FL B-field amplitudes, in tesla.

In the literature, the torque $\boldsymbol{T} \equiv -\frac{1}{|\gamma|}\frac{d\boldsymbol{M}}{dt}$ is expressed either in J.m^{-3}, as here and throughout this textbook, or in A.m^{-1}.s^{-1}, in which case $\boldsymbol{T} \equiv -\frac{d\boldsymbol{M}}{dt}$, resulting in a factor $|\gamma|$ difference is some of the equations depending on the articles and books consulted (see symbols and units page 150).

Note that, the coefficients $t_{\Uparrow(\Downarrow)}$ and $r_{\Uparrow(\Downarrow)}$ account for transmission and reflection of a spin angular momentum $\boldsymbol{s}$ by a ferromagnet of macrospin density $\boldsymbol{S}_M = -\frac{1}{|\gamma|}\boldsymbol{M}$. Equivalently, the coefficients $t_{\uparrow(\downarrow)}$ and $r_{\uparrow(\downarrow)}$ account for transmission and reflection of a spin magnetic moment $\boldsymbol{m} = -|\gamma|\boldsymbol{s}$ by a ferromagnet of magnetization $\boldsymbol{M} = -|\gamma|\boldsymbol{S}_M$. Therefore, $t_{\uparrow(\downarrow)} = t_{\Uparrow(\Downarrow)}$ and $r_{\uparrow(\downarrow)} = r_{\Uparrow(\Downarrow)}$, and $g^{\uparrow\downarrow} = 1 - \left(t_\uparrow t_\downarrow^* + r_\uparrow r_\downarrow^*\right) = 1 - \left(t_\Uparrow t_\Downarrow^* + r_\Uparrow r_\Downarrow^*\right)$.

4.3 STT in CPP Transport Equations

In the previous section, we have described the transfer of angular momentum from the incoming spins to magnetization. We have also detailed how several physical effects, which are combined in a single parameter, the spin mixing conductance, govern the transfer efficiency. The transfer of angular momentum represents a loss of momentum for the spin current, which must be introduced in the diffusion equations of transport. More specifically, it is added in the equation describing the flux of spin current, in the same manner as the loss of angular momentum for the spin current through the spin-flip process was introduced in equation (26).

Taking into account STT, the flux of spin current reads

$$\nabla \cdot \boldsymbol{J}_s = \frac{\sigma^*\left(1-\beta^2\right)}{4el_{sf}^{*\,2}}\boldsymbol{\mu}_s + \frac{\sigma^*}{4eM_S l_{sd}^2}\left(\boldsymbol{M}\times\boldsymbol{\mu}_s\right) \tag{64}$$

with l_{sf}^* the average spin-diffusion length defined in §3.1, l_{sd} the 'exchange' length accounting for the absorption of the transverse angular momentum, and $\sigma^* = \sigma\left(1-\beta^2\right)$.

Basically, equation (64) accounts for the fact that any loss in spin current must be due either to spin-flips, represented by the 1st term, also equation (26) (demonstration in exercise 6.3), or to STT, involving damped precession of the spin around the local magnetization due to *sd* exchange interaction, represented by the 2nd term. In metals, l_{sd} is typically equal to a few nanometers. In the most general case, a generic length (see also §3.5 and §4.5) is defined rather than l_{sd}. It also takes into account precession around the spin-orbit field. In materials with large spin-orbit interactions, like semiconductors, this latter effect is no longer negligible.

We note that we consider here the three-dimensional case because precessions are taken into account. Therefore, in equation (64), the spin current $\boldsymbol{J}_s$ is a tensor, and its divergence and spin accumulation $\boldsymbol{\mu}_s$ are vectors. This case contrasts with the one-dimensional case of equation (26), in which $\boldsymbol{J}_s$ was a vector, and its divergence and the spin accumulation μ_s were scalars.

Typical critical current densities for switching magnetization by STT are $J_{e,\text{crit}} \sim 10^7$ A.cm^{-2}, hence the need for nanometer-size lateral dimensions (figure 39, top-left) – in addition to nm-thick layers, because the layers' thicknesses need to be smaller than l^*_{sf}.

From figure 39, we observe that switching is polarity-dependent and that the positive critical current density is larger than the negative one: $J^+_{e,\text{crit}} > J^-_{e,\text{crit}}$. In fact, for a positive electron flux $J^-_{e,\text{crit}}$, spins travel only once across the N (Cu) layer before acting on the F_{free} layer, whereas for a negative electron flux $J^+_{e,\text{crit}}$, spins travel twice across the N (Cu) layer (figure 39, bottom). As a result, the exponential increase of $J^+_{e,\text{crit}}$ is twice faster than $J^-_{e,\text{crit}}$ (figure 39, top-right).

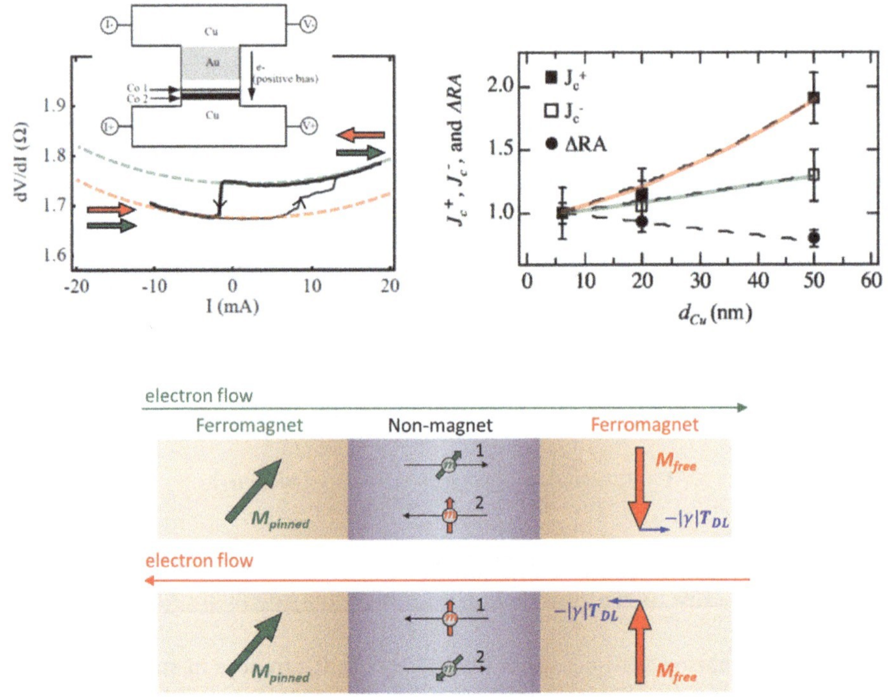

FIG. 39 – (Top-Left) Experimental result showing the current induced STT-switching of the magnetization of the free-ferromagnet, and the corresponding change of resistance due to the GMR effect; inset: schematics of a representative pillar used for STT experiments, making use of a pinned-ferromagnet/non-magnet/free-ferromagnet trilayer, reprinted with permission from the American Physical Society: J. A. Katine *et al.*, *Phys. Rev. Lett.* **84**, 3149 (2000). Copyright 2000. (Top-Right) Dependence of the critical switching current on the thickness of the non-magnet, for two cases: J^+_e when electrons flow from the free- to the pinned-ferromagnet, and J^-_e in the opposite direction. Reprinted with permission from the American Physical Society: F. J. Albert *et al.*, *Phys. Rev. Lett.* **89**, 226802 (2002). Copyright 2002. (Bottom) Illustration of the STT process in the two cases presented, with $d\boldsymbol{M}/dt = -|\gamma|\boldsymbol{T}$.

4.4 STT in Magnetization Dynamics Equations, Reciprocal Spin Pumping

■ The STT terms in magnetization dynamics equations

Naturally, the STT terms can be introduced in the Landau–Lifshitz–Gilbert (LLG) equation, which describes magnetization dynamics [basics about the LLG equation can, for example, be gained in J. M. D. Coey, *Magnetism and Magnetic Materials*, Cambridge University Press (2010)]. The result of adding the STT terms leads to

$$\begin{aligned}\frac{d\boldsymbol{M}}{dt} &= -|\gamma|\boldsymbol{T}_{\text{Larmor}} - |\gamma|\boldsymbol{T}_{\text{Damping}} - |\gamma|\boldsymbol{T}_{\text{DL}} - |\gamma|\boldsymbol{T}_{\text{FL}} \\ \frac{d\boldsymbol{M}}{dt} &= -|\gamma|\boldsymbol{M}\times(\mu_0\boldsymbol{H}_{\text{eff}}) + \frac{\alpha}{M_S}\boldsymbol{M}\times\frac{d\boldsymbol{M}}{dt} + |\gamma|\frac{\tau_{\text{DL}}}{M_S^2}\boldsymbol{M}\times(\boldsymbol{m}\times\boldsymbol{M}) \\ &\quad + |\gamma|\frac{\tau_{\text{FL}}}{M_S}(\boldsymbol{m}\times\boldsymbol{M})\end{aligned} \tag{65}$$

The first term in equation (65) represents the Larmor precession around the effective field $\boldsymbol{H}_{\text{eff}}$. It includes the contributions of the external magnetic field and internal fields due to anisotropy, exchange and dipolar interactions. We recall that the gyromagnetic ratio $\gamma = -\frac{e}{2m_e}g$, with g the Landé g-factor, is negative because of the negative charge $-e$ of the electron, making the precession anticlockwise viewed from above. In the literature, the gyromagnetic ratio $\gamma = -\frac{e}{2m_e}g$ is used either as is in the equations, $\frac{d\boldsymbol{M}}{dt} = \gamma\boldsymbol{M}\times(\mu_0\boldsymbol{H}_{\text{eff}})\ldots$, or in absolute value $|\gamma|$, as throughout this textbook, $\frac{d\boldsymbol{M}}{dt} = -|\gamma|\boldsymbol{M}\times(\mu_0\boldsymbol{H}_{\text{eff}})\ldots$, or replaced by $\gamma_0 = -\gamma\mu_0$, to make it positive and include μ_0 to match the cgs unit system, $\frac{d\boldsymbol{M}}{dt} = -\gamma_0\boldsymbol{M}\times\boldsymbol{H}_{\text{eff}}\ldots$, resulting in a factor -1 or $\pm\mu_0$ difference is some of the equations depending on the articles and books consulted (see symbols and units page 150).

The second term in equation (65) represents damping α, due to internal relaxation ascribed to intraband and interband scattering (exercice 6.5), and to external relaxation, for example, due to spin pumping (exercise 6.7).

Finally, the last two terms represent the DL and FL components of STT, respectively. How the two contributions to STT were named is clear from the LLG equations and corresponding representation (figure 40), because the DL torque is along the damping or anti-damping torque, and the FL torque is along the Larmor or anti-Larmor torque.

Because the FL term is most often negligible in metals (§4.2), we focus on the DL term to estimate the critical current needed to trigger magnetization dynamics (precession or switching, figure 40). The amplitude of the DL torque is expressed as a function of the material's parameters as follows: $\tau_{\text{DL}} = \frac{J_e\hbar\beta}{2eM_Sd_F}$, with d_F the thickness of the magnetic layer, β the polarization, and J_e the amplitude of the charge current density.

The sign of the torque depends on the current direction. We recall that, in regards to the orientation of the torques, the rule is as follows: when an electron of

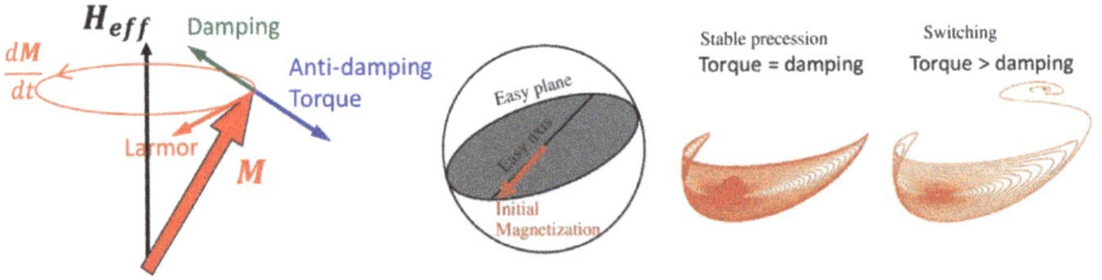

FIG. 40 – (Left) Schematics of magnetization dynamics when the magnetization $\boldsymbol{M}$ is subjected to several torques: the Larmor torque created by the (effective) magnetic field (H_{eff}), the Damping torque due to energy transfer, for example, to electrons, magnons and phonons, and the Anti-Damping component of the spin transfer torque (STT). (Right) Calculations illustrating two cases: stable precession and switching, depending on the relative magnitudes of the torques involved. Reprinted with permission from Elsevier: D. C. Ralph and M. D. Stiles, *J. Mag. Magn. Mat.* **320**, 1190 (2008). Copyright 2008.

magnetic moment density $\boldsymbol{m}$ (or equivalently of spin density $\boldsymbol{s}$ pointing in the opposite direction) flows toward a magnetization $\boldsymbol{M}$, the DL torque orients $\boldsymbol{M}$ toward $\boldsymbol{m}$ (away from $\boldsymbol{s}$).

The critical current density is obtained when the (anti)DL torque compensates the damping torque in each period of oscillation of $\boldsymbol{M}$

$$|\gamma|\tau_{\text{DL,crit}} = 2\pi\frac{\alpha}{dt} = 2\pi\alpha f \tag{66}$$

$$J_{e,\text{crit}} = \frac{e\mu_0 M_S^2 d_F \alpha}{\hbar\beta} \tag{67}$$

The resonance frequency of a ferromagnet is in the GHz range. For an isotropic thin film with a planar magnetization subject to an applied field in the plane, Kittel's law describing the frequency-field relation is: $2\pi f = |\gamma|\mu_0\sqrt{H(H+M_S)}$ [J. M. D. Coey, *Magnetism and Magnetic Materials*, Cambridge University Press (2010)]. For the calculation, we took the frequency corresponding to the zero-field value of the high-field asymptotic dependence: $2\pi f = \frac{|\gamma|\mu_0 M_S}{2}$.

The reasoning is better when considering energies. The critical current density is obtained when the energy lost due to damping is compensated by the energy gained due to antidamping by STT in each period of oscillation of $\boldsymbol{M}$.

The parameters for $Ni_{80}Fe_{20}$ permalloy are $\alpha = 0.008$, $M_S = 0.8\,\text{MA.m}^{-1}$, and $\beta = 0.3$. If we consider $d_F = 3\,\text{nm}$, and the values of the physical constants given page 150, we obtain $J_{e,\text{crit}} \sim 10^7\,\text{A.cm}^{-2}$, indeed one order of magnitude larger compared to the current needed to observe the GMR effect (§3.3). We also get $\tau_{\text{DL}}[\mu T] \sim 40 J_e[10^7\,\text{A.cm}^{-2}]$.

STT is now used in commercial MRAM applications, for writing, when TMR is used for reading. How the fundamental effects described in this book are used in MRAM applications, is detailed in B. Dieny, R. B. Goldfarb, K.-J. Lee (eds) *Introduction to Magnetic Random-Access Memory*, Wiley (2016).

■ STT – spin pumping reciprocity

Spin pumping and spin transfer torque are reciprocal effects [Y. Tserkovnyak *et al.*, *Rev. Mod. Phys.* **77**, 1375 (2005)]. An intuitive picture consists in comparing spin transfer torque to a water flow (spin current) moving the blades of a watermill (magnetization) and spin pumping to moving blades (magnetization) creating a water flow (spin current) (figure 41). The spin pumping effect will be described in detail in exercises 6.6 and 6.7.

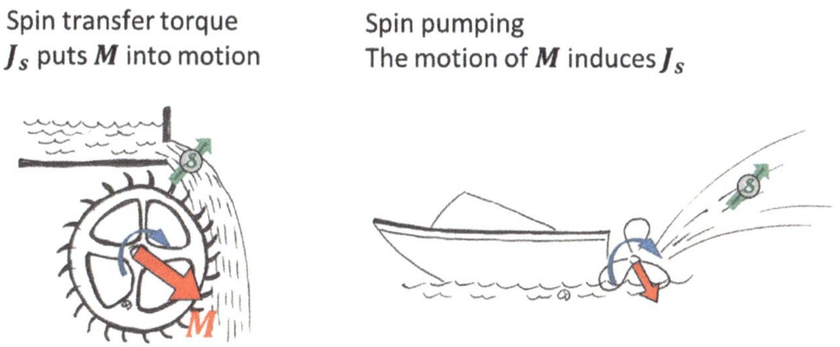

FIG. 41 – Schematics showing (Left) spin transfer torque and (Right) the spin pumping reciprocal effect. The drawings are courtesy of D. Gusakova.

In short for now, the spin pumping effect refers to the ability of a magnetic material, when brought out-of-equilibrium, to generate a spin current ($\boldsymbol{J}_s^0$) at an interface, due to transfer of angular momentum. The technique usually involves inducing resonance in a ferromagnetic (F) spin injector – *e.g.*, a NiFe layer – which is adjacent to a non-magnetic material (N) known as the spin sink – *e.g.*, a Pt layer (figure 42). Spin pumping is therefore treated as a loss of momentum for the spin current, due to transfer of angular momentum at the F/N interface, thus impacting spin accumulation through the addition of a relaxation channel [equation (26)] and magnetization dynamics through the addition of a STT term [equation (65)]. This is similar to what we have described in the previous sections, but here, relaxation and STT are induced by a pure spin current rather than a spin-polarized charge current.

Transport: understanding which material's parameters drive spin pumping (and a subsequent spin-charge conversion) is the object of exercise 6.6. In this exercise, based on the diffusion equation for spin accumulation [equation (27)], we will understand why the spin current propagating in the N layer, along y, reads

$$J_s(y) = J_s^0 \frac{\sinh\left((d_N - y)/l^*_{sf,N}\right)}{\sinh\left(d_N/l^*_{sf,N}\right)} \tag{68}$$

and why this gives rise to a transverse voltage, due to the inverse spin Hall effect (ISHE)

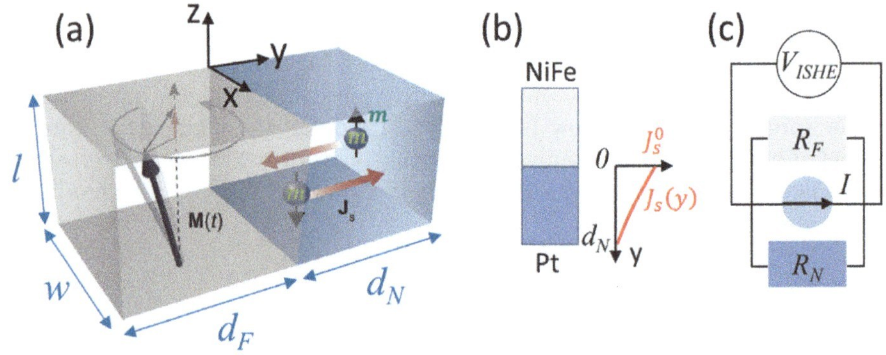

FIG. 42 – (a) Illustration of the spin pumping effect due to sustained out-of-equilibrium magnetization dynamics. (b) Illustration of spin current diffusion across the N layer (here, Pt). (c) Equivalent circuit considering spin current to charge current conversion due to the inverse spin Hall effect in the N layer. Adapted with permission from AIP Publishing: K. Ando *et al.*, *J. Appl. Phys.* **109**, 103913 (2011). Copyright 2011.

$$V_{\text{ISHE}} = \frac{w\theta_{\text{SHE}} l_{sf}^* \tanh\left(d_N/(2l_{sf}^*)\right)}{d_N\sigma_N + d_F\sigma_F} \frac{eg_r^{\uparrow\downarrow}\gamma^2\left(\mu_0 h_{rf}\right)^2}{8\pi\alpha^2\omega} \sin(\theta_M)\overline{\Gamma} \tag{69}$$

with w the width of the layers, θ_{SHE} the spin-Hall angle, h_{rf} the *rf* excitation field used to reach F resonance, and ω the resonance angular frequency. $\overline{\Gamma}$ is a parameter accounting for the trajectory of the magnetization precession in the xy plane. It can be viewed as the elliptic- to circular-trajectory ratio. $g_r^{\uparrow\downarrow}$ is the real part of the spin mixing conductance per unit area per quantum conductance per spin channel, accounting for the ability of the F/N interface and N layer to absorb the spin component along $\boldsymbol{M} \times (\boldsymbol{m} \times \boldsymbol{M})$.

Magnetization dynamics: how spin pumping gives rise to extra damping for magnetization dynamics is the object of exercise 6.7. We will show that the additional torque term due to spin pumping reads [R. Igushi *et al.*, *J. Phys. Soc. Jap.* **86**, 011003 (2017)]

$$-\boldsymbol{T}_{\text{STT}}^{\text{SP}} = \boldsymbol{\nabla} \cdot \left(\frac{\hbar}{2e}\boldsymbol{J}_s\right) = \frac{\alpha_p}{|\gamma| M_S} \boldsymbol{M} \times \frac{d\boldsymbol{M}}{dt} \tag{70}$$

where

$$\alpha_p = \frac{|\gamma|}{M_S d_F} \frac{\hbar}{4\pi} g_{r,\text{eff}}^{\uparrow\downarrow} \tag{71}$$

is an extrinsic damping term showing that spin pumping creates a loss of angular momentum for $\boldsymbol{M}$, with $g_{r,\text{eff}}^{\uparrow\downarrow}$ the real part of the effective spin mixing conductance per unit area per quantum conductance per spin channel, that takes into account back reflections of the spin current at the outer interfaces of the F/N bilayer, compared to $g_r^{\uparrow\downarrow}$ which does not.

■ Note about the general expression of the torque due to a spin or spin-polarized current

Another way of finding the expression of the torque created by a spin current or spin-polarized current on magnetization is shown in figure 43. It consists in looking at the conservation of total magnetic moment (without spin-flip relaxation) by analogy with the conservation of charge [equations (72) and (73), where ρ is the charge density defined in equation (47)]. From this, we get equations (74) and (75). Consequently, the torque created by a flow of spin $\boldsymbol{s}$ angular momentum density $\boldsymbol{Q} = \frac{\hbar}{2e}\boldsymbol{J}_s$ on a magnetization $\boldsymbol{M}$ is $-\boldsymbol{T} = \boldsymbol{\nabla} \cdot \left(\frac{\hbar}{2e}\boldsymbol{J}_s\right)$, with $\boldsymbol{J}_s = -\left(\boldsymbol{j}^{\uparrow} - \boldsymbol{j}^{\downarrow}\right)$, where the ↑ and ↓ symbols refer to the orientation of the moment $\boldsymbol{m}$.

We note that, naturally, the torque created by a flow of spin $\boldsymbol{s}$ angular momentum density $\boldsymbol{Q} = \frac{\hbar}{2e}\boldsymbol{J}_s$ on a macrospin angular momentum density $\boldsymbol{S}_M$ is of opposite sign, as $\boldsymbol{S}_M$ and $\boldsymbol{M}$ are of opposite sign, in agreement with equations (52) and (58) from the toy quantum mechanical model.

$$\iiint \frac{d\rho}{dt}\, dV = -\iint J_e dS \tag{72}$$

$$\frac{d\rho}{dt} = -\boldsymbol{\nabla} \cdot J_e \tag{73}$$

$$\iiint \frac{d\boldsymbol{M}}{dt}\, dV = -\iint \frac{\mu_B g}{2e}\left(\boldsymbol{j}^{\uparrow} - \boldsymbol{j}^{\downarrow}\right) dS \tag{74}$$

$$\frac{d\boldsymbol{M}}{dt} = |\gamma| \boldsymbol{\nabla} \cdot \left(\frac{\hbar}{2e}\boldsymbol{J}_s\right) = -|\gamma| \boldsymbol{T} \tag{75}$$

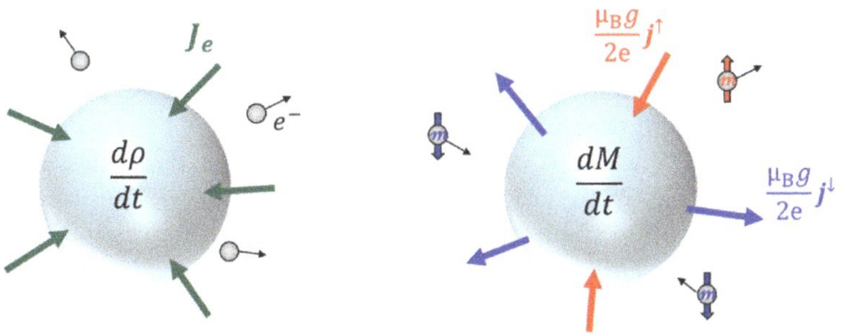

FIG. 43 – Illustration of (Left) charge conservation, and (Right) magnetic moment conservation.

4.5 STT in Magnetic Textures

Up until now, we considered STT in the case of uniform magnetization. In this section, we will deal with the case of non-uniform magnetization in magnetic textures like domain walls (DWs). This part echoes §3.5 in which we discussed DW magnetoresistance.

Taking $\frac{d\boldsymbol{M}}{dt} = -|\gamma|\boldsymbol{T}$, the STT in the case of a DW is expressed as follows

$$-\boldsymbol{T} = \frac{\hbar\beta}{2eM_S^3(1+\xi^2)}\boldsymbol{M}\times[\boldsymbol{M}\times(\boldsymbol{J}_e\cdot\nabla)\boldsymbol{M}] + \frac{\hbar\beta\xi}{2eM_S^2(1+\xi^2)}\boldsymbol{M}\times(\boldsymbol{J}_e\cdot\nabla)\boldsymbol{M} \quad (76)$$

We recall that here, for consistency, β is the polarization of the current, not to be confused with the 'beta term' found in the literature, which takes into account the adiabaticity of the process.

In fact, two contributions are usually considered, depending on the ability of the spins $\boldsymbol{s}$ (or equivalently moments $\boldsymbol{m} = -|\gamma|\boldsymbol{s}$) of the itinerant s-electrons to follow magnetization $\boldsymbol{M}$ (figure 32):

- **the adiabatic (DL) contribution**, when there is perfect tracking, represented by the 1st term in equation (76), and
- **the non-adiabatic (FL) contribution**, when there is mistracking; 2nd term in equation (76).

The adiabatic contribution is very similar to the mechanism described in §4.2, in which the spin current or flux of spin density transfers angular momentum to magnetization due to sd-interactions, therefore generating a predominant DL torque. In the non-adiabatic contribution, spin accumulation takes place due to spin-mistracking. It then precesses around the magnetization and generates a FL torque upon relaxation.

We recall (§3.5) that the process of spin-mistracking is driven by the spin-mistracking parameter ξ. In low spin-orbit materials, $\xi = \frac{\tau_{sd}}{\tau_{sf}} = \left(\frac{l_{sd}}{l_{sf}}\right)^2$, with $l_{sd} = \sqrt{D\tau_{sd}}$ and $l_{sf} = \sqrt{D\tau_{sf}}$. The precession of itinerant electrons around the spin-orbit field enhances ξ. In weakly spin-orbit-coupled dense-moment F metals like Co, $\xi \ll 1$, and the DL term dominates the FL one: DL $\gg$ FL. In contrast, in strongly spin-orbit-coupled dilute-moment F semiconductors, like (Ga,Mn)As, $\xi \gg 1$, and the FL term dominates: FL $\gg$ DL.

In the same way that STT can trigger magnetization dynamics in a uniform system, it can also move magnetic textures. The torques contributions distort the initial Bloch configuration (figure 44), therefore redistributing the initial conditions for the calculation of $\boldsymbol{T}$. This point underlines that there is constant feedback between $\nabla\boldsymbol{M}$ and $\boldsymbol{T}$.

A possible application of STT in magnetic textures is the racetrack memory (figure 45) [S. S. P. Parkin *et al.*, *Science* **320**, 190 (2008)], in which it is possible to store information in a 3-dimensional way, unlike current storage technologies. The interested reader of 3-dimensional spintronics applications may consult focused reviews like A. Fernández-Pacheco *et al.*, *Nat. Commun.* **8**, 15756 (2017).

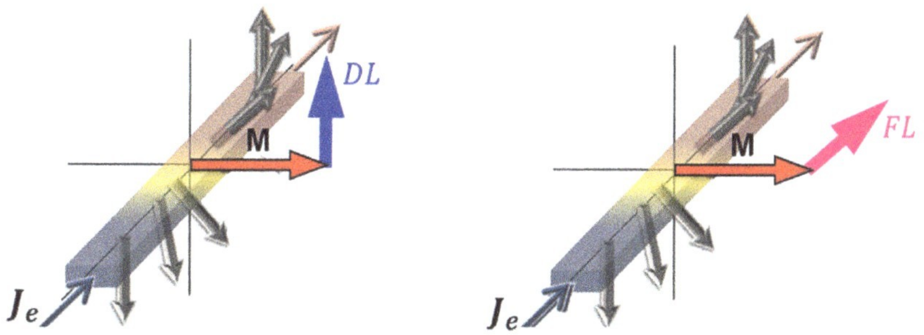

FIG. 44 – Schematics illustrating how current-induced (Left) damping-like (DL) and (Right) field-like (FL) contributions to $d\boldsymbol{M}/dt$ distort and move a Bloch magnetic domain wall. Adapted from T. Janda *et al.*, *Nat. Commun.* **8**, 15226 (2016) – CC BY license.

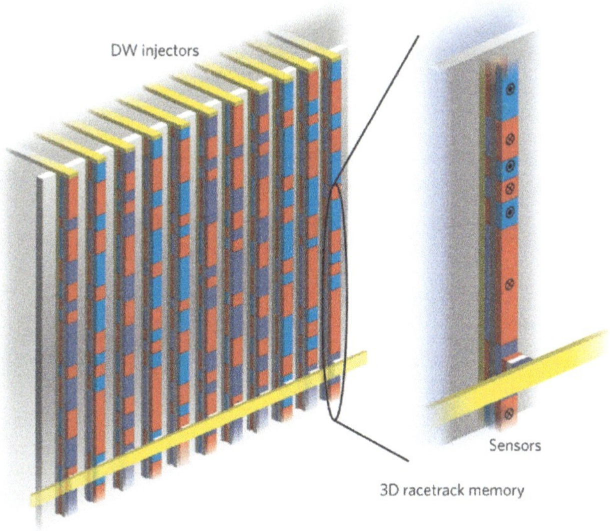

FIG. 45 – Schematics illustrating the principle of racetrack memory. Reprinted with permission from Springer Nature: S. S. P. Parkin *et al.*, *Nat. Nanotech.* **10**, 195 (2015). Copyright 2015.

4.6 Magnetoelectronic Circuit Theory

In this section, we want to briefly mention the magnetoelectronic circuit theory of transport [A. Braatas, G. E. W. Bauer, and P. J. Kelly, *Phys. Rep.* **427**, 157 (2006)]. It is a generalization of circuit theory to non-collinear magnetization. The findings of this theory are now commonly used in the field of spintronics, be it related to CPP-GMR, STT, spin pumping, or other effects like the spin Hall magnetoresistance (SMR), which is treated in exercise 6.8. Because CPP-GMR, STT, and spin pumping are companion effects, it is possible to measure the same type of parameters, like the spin mixing conductance $G^{\uparrow\downarrow}$ and the spin diffusion length l^*_{sf}, in different ways. While this section is intended for advanced readers, the comparison of the findings of the model with the toy model for STT discussed in §4.2 is within reach of all readers.

In this theory, the spatial dependence of spin accumulation is disregarded. Spin-accumulation is reduced to a vector-potential $\boldsymbol{V}_{s,N}$ and the N layer is reduced to a node with a set number of net spins that have accumulated $Ne|\boldsymbol{V}_{s,N}|$ (figure 46). The initial model is therefore appropriate when interfacial resistances prevail. We note that, in this book, for the sake of consistency, $\boldsymbol{V}_{s,N}$ refers to the vector potential representing electrons with spin $\boldsymbol{s}$ in the N layer, meaning that in the F layer, $\boldsymbol{V}_{s,F} = \hat{\boldsymbol{s}}\, V_{s,F} = -\widehat{\boldsymbol{m}}\, V_{s,F}$.

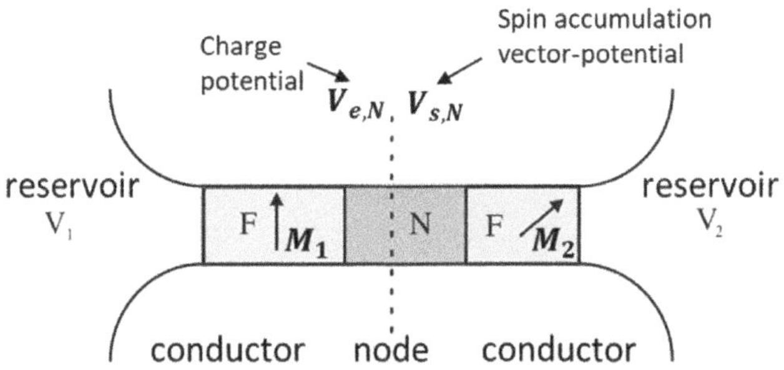

FIG. 46 – Schematics of the equivalent circuit used to model STT in the frame of the magnetoelectronic circuit theory of transport. Adapted with permission from Elsevier: A. Braatas, G. E. W. Bauer, and P. J. Kelly, *Phys. Rep.* **427**, 157 (2006). Copyright 2006.

The transverse spin $\boldsymbol{s}$ current and current density, at the interface, arriving from the layer N to the layer F_1 (figure 46) have the form

$$\boldsymbol{J}_s A = 2\left[\frac{G_r^{\uparrow\downarrow}}{M_S^2}\boldsymbol{M}_1 \times \left(\boldsymbol{M}_1 \times \boldsymbol{V}_{s,N}\right) + \frac{G_i^{\uparrow\downarrow}}{M_S}\boldsymbol{M}_1 \times \boldsymbol{V}_{s,N}\right] \tag{77}$$

$$\boldsymbol{J}_s = \frac{2G_r^{\uparrow\downarrow}}{AM_S^2}\boldsymbol{M} \times \left(\boldsymbol{M} \times \frac{\boldsymbol{\mu}_s}{e}\right) + \frac{2G_i^{\uparrow\downarrow}}{AM_S}\boldsymbol{M} \times \frac{\boldsymbol{\mu}_s}{e} \tag{78}$$

where $G_r^{\uparrow\downarrow} = \mathcal{R}e\left(G^{\uparrow\downarrow}\right)$ and $G_i^{\uparrow\downarrow} = \Im m\left(G^{\uparrow\downarrow}\right)$ are the real and imaginary parts of the spin mixing conductance, accounting for the spin current that is polarized along the two axes transverse to $\boldsymbol{M}$, and that is therefore absorbed by the magnetic material. We recall that the absorption of the transverse angular momentum creates a torque. Given the transverse direction to which they refer, $G_r^{\uparrow\downarrow}$ results in the damping-like torque and $G_i^{\uparrow\downarrow}$ results in the field-like torque. In the literature, the spin mixing conductance is defined either as $G^{\uparrow\downarrow}$, in siemens as here and throughout this textbook, or as $G^{\uparrow\downarrow}/A$, in which case it is in units of S.m^{-2}, resulting in a factor A difference in some of the equations depending on the articles and books consulted. We also recall that the spin mixing conductance is most often used per unit area A per quantum conductance per spin channel $G_0/2$: $g^{\uparrow\downarrow} = \frac{G^{\uparrow\downarrow}}{AG_0/2}$. It is customary to incorrectly use the term spin mixing conductance for $g^{\uparrow\downarrow}$, although its unit is m^{-2} (see symbols and units page 150).

The torque acting on F_1, due to absorption of the transverse spin current is then

$$-\boldsymbol{T} = \nabla \cdot \left(\frac{\hbar}{2e}\boldsymbol{J}_s\right) = \frac{\hbar}{e^2 V}\left[\frac{G_r^{\uparrow\downarrow}}{M_S^2}\boldsymbol{M} \times (\boldsymbol{M} \times \boldsymbol{\mu}_s) + \frac{G_i^{\uparrow\downarrow}}{M_S}\boldsymbol{M} \times \boldsymbol{\mu}_s\right] \tag{79}$$

where

$$G^{\uparrow\downarrow} = \frac{e^2}{h}\sum_{n\in N}\left[1 - \sum_{m\in N} r_{\Uparrow}^{nm}\left(r_{\Downarrow}^{nm}\right)^*\right] \tag{80}$$

The integers n, m label what are called the propagating modes or scattering channels. They actually relate to the wave numbers $k_{n,m}$ at the Fermi level. The elements $r_{\Uparrow(\Downarrow)}^{nm}$ are the reflection coefficients for a transverse mode m with spin $\Uparrow$ ($\Downarrow$) at the N side reflected to a transverse mode n with spin $\Uparrow$ ($\Downarrow$).

We recall that the coefficients $t_{\Uparrow(\Downarrow)}$ and $r_{\Uparrow(\Downarrow)}$ account for transmission and reflection of a spin $\boldsymbol{s}$ by a ferromagnet of macrospin density $\boldsymbol{S}_M = -\frac{1}{|\gamma|}\boldsymbol{M}$. Equivalently, the coefficients $t_{\uparrow(\downarrow)}$ and $r_{\uparrow(\downarrow)}$ account for transmission and reflection of a moment $\boldsymbol{m} = -|\gamma|\boldsymbol{s}$ by a ferromagnet of magnetization $\boldsymbol{M} = -|\gamma|\boldsymbol{S}_M$. Therefore, $t_{\uparrow(\downarrow)} = t_{\Uparrow(\Downarrow)}$ and $r_{\uparrow(\downarrow)} = r_{\Uparrow(\Downarrow)}$, and $G^{\uparrow\downarrow} = \frac{e^2}{h}\sum_{n\in N}\left[1 - \sum_{m\in N} r_{\Uparrow}^{nm}\left(r_{\Downarrow}^{nm}\right)^*\right] = \frac{e^2}{h}\sum_{n\in N}\left[1 - \sum_{m\in N} r_{\uparrow}^{nm}\left(r_{\downarrow}^{nm}\right)^*\right]$.

Combining equations (79) and (80), it is possible to express the spin transfer torque in the magnetoelectronic circuit theory

$$\begin{aligned} -\boldsymbol{T} = \frac{1}{2\pi V}\Bigg[&\left(N - \mathcal{R}e\left(\sum_{m\in N} r_{\Uparrow}^{nm}\left(r_{\Downarrow}^{nm}\right)^*\right)\right)\widehat{\boldsymbol{M}} \times \left(\widehat{\boldsymbol{M}} \times \boldsymbol{\mu}_s\right) \\ &- \mathfrak{Im}\left(\sum_{m\in N} r_{\Uparrow}^{nm}\left(r_{\Downarrow}^{nm}\right)^*\right)\widehat{\boldsymbol{M}} \times \boldsymbol{\mu}_s\Bigg] \end{aligned} \tag{81}$$

where N represents the number of scattering channels in the N layer.

This expression is very similar to that obtained with the toy quantum mechanical model [equation (57)], with its one type of electron wave – which can be refined with a summation over the Fermi surface. For the comparison, we recall that the notations $\boldsymbol{M}$ and $\boldsymbol{S}_M$ refer to the magnetization and macrospin density of the ferromagnet *receiving* the flux of angular momentum, and $\boldsymbol{\mu}_s$, $\boldsymbol{s}$, and $\boldsymbol{m}$ to the spin $\boldsymbol{s}$ accumulation, spin $\boldsymbol{s}$ current density, and magnetic moment $\boldsymbol{m}$ density of the incoming electrons *bringing* a flux of angular momentum (see symbols and units page 150).

4.7 Spin-Orbit Torques – SOT

In this section, we now turn to another type of torque that exists in spintronics: spin-orbit torque (SOT). We describe several types of SOT and show how they result from the transfer of angular momentum from the electron current to magnetization, through interactions between lattice and electron orbital (crystal field potential), electron orbital and spin (spin-orbit coupling), and spin and magnetization (*sd*-exchange interactions).

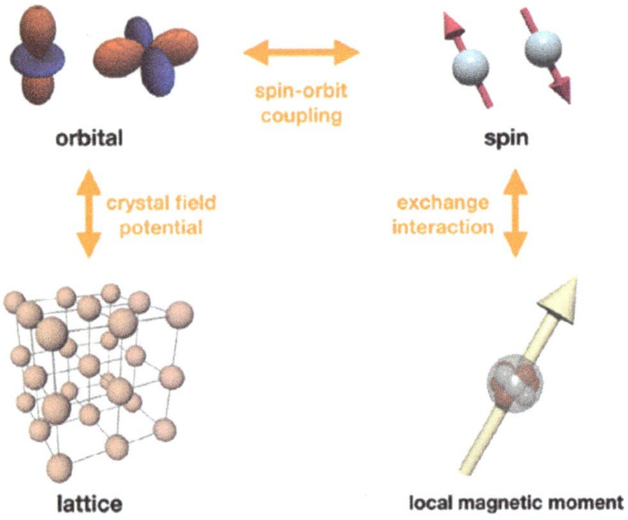

FIG. 47 – Schemes of four reservoirs of angular momentum in crystals – spin and orbital of the electron, the crystal lattice, and the local magnetic moment – and the sources of interactions and transfer/interconversion between them – (*sd*)-exchange interactions (see also chapter 4), spin-orbit coupling (§2.6), and crystal field potential due to Coulomb interactions (chapter 5). From D. Go *et al.*, *Phys. Rev. Res.* **2**, 033401 (2020) – CC BY license.

The concept of SOT refers to a torque created by an electric current on a magnetization, the origin of which involves a spin-orbit coupling mechanism. More generally, this type of torque involves interaction processes between different sources of angular momentum in crystals (figure 47):

- ***sd*-exchange interactions** between magnetic moments due to localized electrons and the spin of the itinerant conduction electrons (see also chapter 4),
- **spin-orbit coupling**, between the spin and orbital angular momenta of the itinerant electrons (see also §2.6),
- **Coulomb interactions (crystal field potential)**, coupling the lattice and the orbital angular momentum of the itinerant electrons. This interaction has given rise to a recent research subfield called orbital spintronics [D. Go *et al.*, *Europhys. Lett.* **135**,

37001 (2021)], in which no spin-orbit coupling is initially required to trigger the spintronic effect (see also chapter 5). In the case of torque, spin-orbit coupling is needed in a second step, to couple the orbital angular momentum and the magnetization to be set in motion. This type of torque is also classified as an orbital torque.

SOT is most often found in non-magnet(N)/ferromagnet(F) bilayers for a current in-plane (CIP) configuration. In contrast to the STT structures described in §4.1–4.6, SOT structures do not require an additional F-layer to create a spin-polarized current (*i.e.*, a flux of angular momentum). In SOT structures, not only spin currents (flux of spin angular momentum, $\boldsymbol{J_s}$), but also non-equilibrium moment densities $\boldsymbol{m}$ (equivalently spin densities $\boldsymbol{s} = -\boldsymbol{m}/|\gamma|$) are generated in the N or F layer, due to spin-orbit coupling.

Spin-orbit effects can be readily understood by considering the motion of an electron in a potential gradient $\nabla\phi$ and how fields are transformed between inertial frames. The net electric field, created by the potential gradient becomes a magnetic field in the electron's rest frame $\boldsymbol{H}_{SO}(\boldsymbol{k})$. This momentum($k$)-dependent magnetic field couples to the magnetic moment of the electron $\boldsymbol{m}$, through a Zeeman term, $-\boldsymbol{m}\cdot\mu_0\boldsymbol{H}_{SO}(\boldsymbol{k})$. In magnetic materials, this interaction interconnects the direction of electron flow and the magnetic order parameter. It is said that the spin angular momentum is locked to its wavevector. Atomic spin-orbit interactions, where ϕ is the potential with lattice periodicity, can be distinguished from effective spin-orbit interactions, which result from combining the atomic spin-orbit interaction with particular crystal or multilayer systems. In crystals or structures lacking inversion symmetry, the k-dependent magnetic field becomes odd in momentum, resulting, for example, in Dresselhaus or Rashba fields (§5.4). [Text adapted from V. Baltz *et al.*, *Rev. Mod. Phys.* **90**, 015005 (2018)].

Several mechanisms have been identified as being at the origin of spin-orbit torque, including the bulk spin Hall effect (SHE) and bulk or interfacial inverse spin galvanic effects (iSGE), like the Rashba–Edelstein effect (REE) (figure 48). In prototypical models, beyond the origin of the effects, the main difference between SHE and iSGE is that the former generates a flux of angular momentum between the N and F layers, whereas the latter generates a non-equilibrium spin density at the N/F interface. As a result, SHE produces a damping-like torque, through absorption of the flux of angular momentum by the F layer, similar to STT (§4.1–4.6), and iSGE produces a field-like torque due to a direct *sd*-exchange mediated by conduction electrons in the F layer, which impinges on and is reflected at the interface. In refined models, and in practice, each torque has both a damping-like and a field-like component, whether induced by SHE or iSGE. They are of the form

$$-\boldsymbol{T} = \frac{\tau_{\rm DL}}{M_S^2}\boldsymbol{M}\times(\boldsymbol{m}\times\boldsymbol{M}) + \frac{\tau_{\rm FL}}{M_S}\boldsymbol{m}\times\boldsymbol{M} \tag{82}$$

where $\boldsymbol{m} \parallel \boldsymbol{J}_e \times \widehat{\boldsymbol{u}}$ is the non-equilibrium moment density, with $\widehat{\boldsymbol{u}}$ a unit vector determined by the symmetry of the system. In the case of a magnetic multilayer perpendicular to $\widehat{z}$, and experiencing SHE or REE as depicted, $\widehat{\boldsymbol{u}} = \widehat{z}$. Two types of these SOT are detailed below.

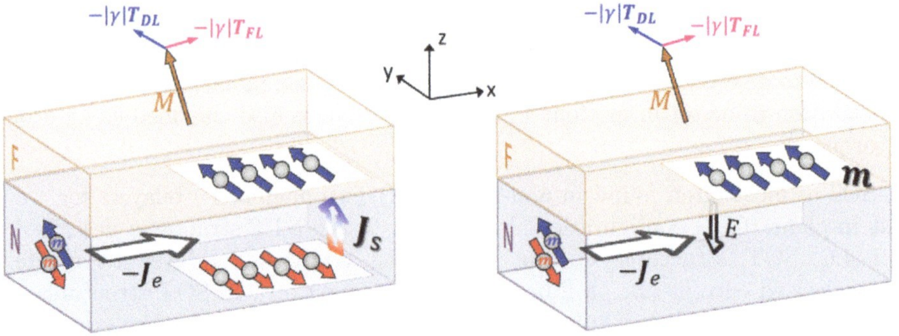

FIG. 48 – Schematics of various origins of the CIP-induced spin-orbit torque in a non-magnet (N)/ferromagnet(F) bilayer. (Left) Spin Hall effect (SHE) takes place in the bulk of the N (heavy metal) layer and creates a flux of spin $\boldsymbol{s}$ angular momentum $\boldsymbol{J}_s$ across the interface. (Right) Inverse spin galvanic Rashba–Edelstein effect (REE) arises at the interface with the N (heavy metal) and creates a non-equilibrium moment density $\boldsymbol{m}$ (equivalently a non-equilibrium spin density $\boldsymbol{s} = -\boldsymbol{m}/|\gamma|$) at the interface.

■ **Bulk SHE-induced torque** (figure 48, left)

Due to bulk spin-orbit coupling in a metallic N layer, electrons of opposite spin scatter in opposite directions [N. Nagaosa *et al.*, *Rev. Mod. Phys.* **82**, 1539 (2010)]. This effect is called the spin Hall effect (SHE), which we admit for now. Thus, a longitudinal charge current is transformed into a transverse spin current in the N and diffuses in the F. This flux of angular momentum creates a torque on the magnetization of the F. The torque is thus similar to that described in §4.1–4.6, although the origin of the flux of angular momentum differs. Therefore, in the case of a pure SHE-induced spin torque, the torque itself is of the STT type. Exercises 6.6 and 6.8 on ISHE and SMR, respectively, give examples of similar physics and how similar torques can be calculated (see also figure 56 and corresponding text).

<u>Damping-like component of the bulk SHE-induced torque.</u>
For $G_i^{\uparrow\downarrow} = 0$ [equations (60)–(63)], the SHE-induced torque takes the form:

$$-\boldsymbol{T}_{\mathrm{STT}}^{\mathrm{SHE}} = \frac{\alpha_{\mathrm{MOD}}^{\mathrm{SHE}}}{|\gamma| M_S} \boldsymbol{M} \times \frac{d\boldsymbol{M}}{dt} = \frac{\tau_{\mathrm{DL}}}{M_S^2} \boldsymbol{M} \times (\boldsymbol{m} \times \boldsymbol{M}) \tag{83}$$

From this equation, we see that $\boldsymbol{T}_{\mathrm{STT}}^{\mathrm{SHE}}$ is damping-like (same as in §4.1–4.6) and thus creates an additional damping term $\alpha_{\mathrm{MOD}}^{\mathrm{SHE}}$. This type of torque enables modulation of the F's damping [K. Ando *et al.*, *Phys. Rev. Lett.* **101**, 036601 (2008)].

Linearizing the LLG equation for magnetization dynamics [equation (65)], it is possible to show that

$$\alpha_{\mathrm{MOD}}^{\mathrm{SHE}} = \frac{1}{\left(H_{\mathrm{ext}} + \frac{M_{\mathrm{eff}}}{2}\right)\mu_0 M_S d_F} \sin(\theta) \frac{\hbar}{2e} J_s^0 \tag{84}$$

The prefactor before $\sin(\theta)\frac{\hbar}{2e}J_s^0$ accounts for the fact that the trajectory of the magnetization is hindered by effective fields, including anisotropy and applied field. The $\sin(\theta)$-dependence, where θ is the angle between $\boldsymbol{J}_e$ and $\boldsymbol{M}$, results from the vector product at the origin of the torque.

Since the spin accumulation and the corresponding transverse spin current $\boldsymbol{J}_s^0$ arise from the longitudinal charge current $\boldsymbol{J}_e$, through the SHE process, we have $J_s^0 = \theta_{\mathrm{SHE}}\eta J_e$. The spin-orbit mechanism is included in the spin Hall angle θ_{SHE} (see figure 56 and corresponding text, also exercise 6.6).

The parameter η is the result of spin accumulation. It takes into account: (i) spin flip process in the non-magnetic layer, *i.e.*, that only a fraction of the converted spin current matters as the conversion far from the interface will be depolarized by the time it reaches the interface, and (ii) the efficiency of the transfer of angular momentum, through the spin mixing conductance (per unit area per spin channel) $g^{\uparrow\downarrow} = G^{\uparrow\downarrow}/(AG_0/2)$. Based on the drift-diffusion equation for spin accumulation [equation (16) for $z = 0$ in Y. T. Chen *et al.*, *Phys. Rev. B* **87**, 144411 (2013), see also exercise 6.8], and considering that the spin accumulation in the F layer is negligible, and that $g_i^{\uparrow\downarrow} = 0$ (§4.2) it can be shown that $\eta = g_r^{\uparrow\downarrow}\tanh\left(\frac{d_N}{2l^*_{sf,N}}\right)/\left[\frac{\sigma_N}{2G_0 l^*_{sf,N}} + g_r^{\uparrow\downarrow}\coth\left(\frac{d_N}{l^*_{sf,N}}\right)\right]$.

The additional damping due to SHE-induced STT is thus

$$\alpha_{\mathrm{MOD}}^{\mathrm{SHE}} = \frac{\sin(\theta)}{\left(H_{\mathrm{ext}} + \frac{M_{\mathrm{eff}}}{2}\right)\mu_0 M_S d_F} \frac{\hbar}{2e} \theta_{\mathrm{SHE}} J_e \frac{g_r^{\uparrow\downarrow}\tanh\left(\frac{d_N}{2l^*_{sf,N}}\right)}{\frac{\sigma_N}{2G_0 l^*_{sf,N}} + g_r^{\uparrow\downarrow}\coth\left(\frac{d_N}{l^*_{sf,N}}\right)} \tag{85}$$

Experimentally, it is most often possible to measure $g_{r,\mathrm{eff}}^{\uparrow\downarrow}$ rather than $g_r^{\uparrow\downarrow}$, for example, from the additional damping due to spin pumping [equation (71)]. How those two parameters are related is the object of exercise 6.7. We recall that $g_{r,\mathrm{eff}}^{\uparrow\downarrow}$ takes into account back reflections of the spin current at the outer interfaces of the F/N bilayer.

Field-like component of the bulk SHE-induced torque.

A field-like contribution to the SHE-induced torque appears when $G_i^{\uparrow\downarrow} \neq 0$, *i.e.*, when considering the spin current that is perpendicular to both the magnetization and the spin accumulation, similar to what was discussed in the section on STT §4.1–4.6.

■ **Interface REE-induced torque** (figure 48, right)

Field-like component of the interface REE-induced torque.

We remind that in systems lacking bulk or interfacial inversion symmetry, the spin-orbit coupling becomes odd in momentum. Examples are linear and cubic Dresselhaus spin-orbit coupling in strained Zinc-Blende semiconductors, Rashba spin-orbit coupling in bulk Wurtzite and at interfaces between dissimilar materials as well as cubic Rashba spin-orbit coupling in oxide heterostructures. The main result of such a type of spin-orbit coupling is that the spin angular momentum is locked perpendicular to the momentum $\boldsymbol{k}$, as shown in figure 49, left and middle for the case of the interfacial Rashba coupling.

A naïve yet pedagogical derivation of the Rashba Hamiltonian $\mathcal{H}_R$ considers that, due to the Lorentz transformation $\mu_0 \boldsymbol{H}_{SO} = -\frac{(\boldsymbol{v}\times\boldsymbol{E})}{c^2}$, electrons experience a magnetic field $\boldsymbol{H}_{SO}$ in their reference frame when moving in the electric field $\boldsymbol{E} = E_z\hat{\boldsymbol{z}}$ created by $\mathcal{P}$-breaking at the surface (figure 48, left). The corresponding Hamiltonian is $\mathcal{H}_R = -\boldsymbol{m}\cdot\mu_0\boldsymbol{H}_{SO} = \frac{g\mu_B}{2c^2}(\boldsymbol{v}\times\boldsymbol{E})\cdot\boldsymbol{\sigma} = \frac{g\mu_B E_0}{2c^2}(\boldsymbol{v}\times\boldsymbol{e}_z)\cdot\boldsymbol{\sigma} = \alpha_R(\boldsymbol{k}\times\boldsymbol{\sigma})\cdot\hat{\boldsymbol{z}}$, where $\alpha_R = \frac{g\mu_B E_0 \hbar}{2m_e c^2}$ is the Rashba parameter. In a refined model, the Dirac gap at play in the naïve model $m_e c^2 \sim 1\,\text{MeV}$ is replaced by an energy term of $\sim 1\,\text{eV}$, which results from splittings in the energy bands, and $\alpha_R \sim 0.1 - 1\,\text{eV.Å}$.

At equilibrium, when $\boldsymbol{J}_e = 0$, the moment density is zero (figure 49, left and middle). In contrast, a non-equilibrium (Edelstein) moment density $\boldsymbol{m} \propto \boldsymbol{J}_e \times \hat{\boldsymbol{u}} \propto \alpha_R\hat{\boldsymbol{y}}$ builds up on the application of a current $\boldsymbol{J}_e = J_x\hat{\boldsymbol{x}}$ (figure 49, right). This moment density is proportional to the Rashba spin-orbit coupling parameter α_R. It leads to a field-like torque of the form $-\boldsymbol{T}_{\text{SOT}}^{\text{REE}} = \frac{\tau_{\text{FL}}}{M_S}\boldsymbol{m}\times\boldsymbol{M}$, due to direct *sd*-exchange mediated by conduction electrons in the F layer [A. Manchon and S. Zhang, *Phys. Rev. B* **78**, 212405 (2008)].

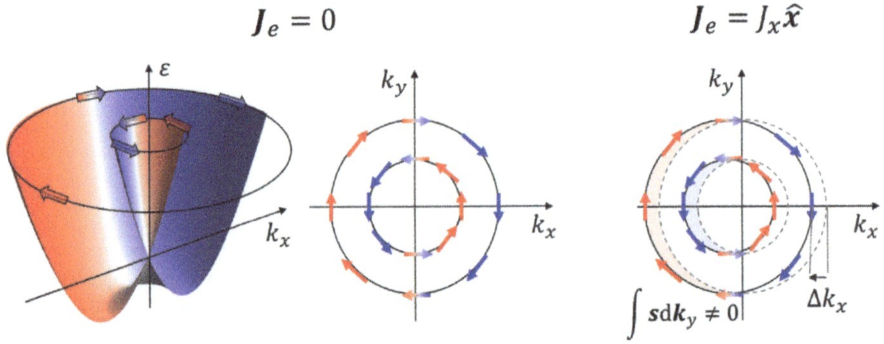

FIG. 49 – Illustration of the Rashba–Edelstein effect (REE). (Left and Middle) 3D-view and cross sections at the Fermi level of the band structures for spin(⇓) and spin(⇑) electrons at equilibrium, for $\boldsymbol{J}_e = 0$, corresponding to spin-orbit induced parity symmetry breaking of the Rashba type, resulting in angular momentum locking. The Hamiltonian reads $\mathcal{H} = \hbar^2\boldsymbol{k}^2/(2m_e) + \alpha_R(\boldsymbol{k}\times\boldsymbol{\sigma})\cdot\hat{\boldsymbol{z}}$ (§5.4). (Right) Edelstein effect in which a non-equilibrium excess spin density $\boldsymbol{s}$ (equivalently moment density $\boldsymbol{m} = -|\gamma|\boldsymbol{s}$) along $-\hat{\boldsymbol{y}}$ ($\hat{\boldsymbol{y}}$) builds up for $\boldsymbol{J}_e = J_x\hat{\boldsymbol{x}} \neq 0$.

Damping-like component of the interface REE-induced torque.

The ideal picture described above considers a two-dimensional spin density, localized at the interface. In practice, spins may accumulate in three dimensions instead. As a result, the spin accumulation can diffuse from the N to the F and thus create a flow of angular momentum, whose absorption by the F layer generates a DL torque of the spin transfer type (§4.1–4.6). Another source of DL torque may come intrinsically from the lattice and can be formulated in the language of the Berry curvature (chapter 5). It is described intuitively by considering the spin-dependent (due to angular momentum locking) electron spin dynamics governed by $\boldsymbol{H}_{SO}(\boldsymbol{k})$ during the acceleration phase, between the scattering events [H. Kurebayashi *et al.*, *Nature Nanotech.* **9**, 211 (2014)].

It is now recognized that SOTs can be classified according to whether the torque originates from the CIP electric current flowing in the N or F, or whether it is due to spin-orbit coupling in the N or F layer [V. P. Amin and M. D. Stiles, *Phys. Rev. B* **94**, 104420 (2016)]. We recall that the SHE-induced torque originates from the electric current flowing in the N, whereas the REE-induced torque originates from the current flowing in the F. In both cases, the torque is due to spin-orbit coupling in the N, in the bulk for the SHE, and at the interface for the REE. The contribution of spin-orbit coupling in the F-layer can also create torques. Because of their common origin, by analogy with the anomalous Hall effect, they are called AHE-induced torques. We will also see in chapter 5 that intrinsic effects may give rise to the so-called spontaneous anomalous Hall effect (SAHE) which can also give rise to SAHE-induced torques. In this last case, the effects arise from the spin-lattice asymmetry, sometimes independent of spin-orbit interactions, and the spin-orbit interaction is only involved in coupling the spin and the crystal lattice, as a perturbation. Other torques, including orbital Hall-effect-induced torques, may also contribute to some structures. Interested readers may consult for example the focused articles by D. Go *et al.*, *Phys. Rev. Res.* **2**, 033401 (2020), and A. Manchon *et al.*, *Rev. Mod. Phys.* **91**, 035004 (2019).

We note that the torques discussed above can contribute concomitantly. Examples for the case of Fe/W and Ni/W bilayers are in D. Go *et al.*, *Phys. Rev. Res.* **2**, 033401 (2020). Using insulating or conducting stacks, or varying the layers' thicknesses may allow a particular type of torque to be selected. We also note that some of these torques can also be encountered when the F is a metal, an insulator, a topological insulator, a 2D material, and if it is replaced by some ferrimagnets or antiferromagnets.

A standard technique to measure spin-orbit torques is based on the harmonic analysis of the periodic Hall voltage following the application of a periodic charge current. This technique is introduced in exercise 6.9.

■ SOT in magnetic textures: the example of the skyrmion Hall effect

Similarly to STT (§4.1–4.6), and thus already discussed in detail, SOT terms can be added to the transport and magnetization dynamics equations and influence the motion of magnetization, be it uniform or textured [D. Go *et al.*, *Phys. Rev. Res.* **2**, 033401 (2020); A. Manchon *et al.*, *Rev. Mod. Phys.* **91**, 035004 (2019)]. To mention just one, a striking example is the SHE-induced torque motion of a skyrmion in an F layer adjacent to a N layer, in which the effect known as the skyrmion Hall effect deviates the skyrmion laterally from the electron flow.

Magnetic skyrmions are nanoscale chiral spin textures that are categorized based on their topological index $Q = \frac{1}{4\pi}\int \widehat{\boldsymbol{M}} \cdot \left(\partial_x \widehat{\boldsymbol{M}} \times \partial_y \widehat{\boldsymbol{M}}\right) dxdy$

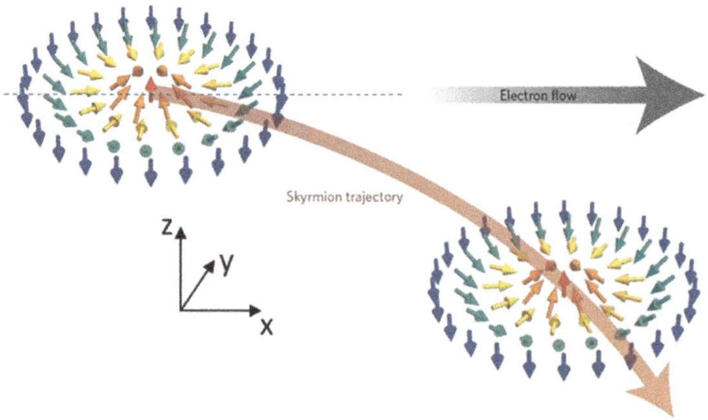

FIG. 50 – Illustration of the SHE-induced torque motion of a skyrmion, in which the skyrmion Hall effect deviates the skyrmion laterally from the electron flow. Reprinted with permission from Springer Nature: G. Chen, *Nat. Phys.* **13**, 112 (2017). Copyright 2017.

[A. Soumyanarayanan *et al.*, *Nature* **539**, 509 (2016)]. As an example, a Néel-type skyrmion is shown in figure 50. By considering a steady and rigid behavior, the in-plane (x, y) motion of the center of mass of a skyrmion with velocity $\boldsymbol{v}$ is best described by a modified Thiele equation (equation 86). It is derived from the known LLG equation (§4.4) in which transport-specific effects are accounted for, such as the damping contribution of the electron spin due to coupling to the skyrmion by the Berry phase and the forces on the electrons arising from the Berry phase they pick up when their spin adiabatically follows the topology of the moments in the skyrmion (see chapter 5 for a description of the Berry phase and the topological Hall effect, companion to the skyrmion Hall effect). Detailed calculations showing how the transport-specific effects for a skyrmion are included in the LLG equation can, for example, be found in K. Everschor *et al.*, *Phys. Rev. B* **86**, 054432 (2012). The Thiele equation thus writes:

$$\boldsymbol{G} \times \boldsymbol{v} - \alpha\overline{\overline{D}} \cdot \boldsymbol{v} + \boldsymbol{F} = 0 \tag{86}$$

The first term in the equation describes the topological Magnus force. For a Néel-type two-dimensional skyrmion, one gets $\boldsymbol{G} = -\frac{M_S d_F}{|\gamma|} 4\pi Q\widehat{\boldsymbol{z}}$, known as the gyromagnetic coupling vector, indeed related to the skyrmion topological index Q. The second term represents the dissipative force. It is governed by the dyadic tensor $\overline{\overline{D}}$, with $D_{ij} = \frac{M_S d_F}{|\gamma|} \int \left(\partial_i \widehat{\boldsymbol{M}} \cdot \partial_j \widehat{\boldsymbol{M}} \right) didj$. Here, $D = D_{xx} = D_{yy}$ and $D_{xy} = D_{yx} = 0$. The last term is the force at the origin of the motion. For the case of a SHE-induced torque, $\boldsymbol{F}$ relates to the force on the electrons arising from Berry phase inherent to the magnetic topology of the skyrmion. It writes $\boldsymbol{F} = 4\pi\overline{\overline{B}} \cdot \boldsymbol{J}_e$, where $\overline{\overline{B}}$ is a tensor related to the emergent topological magnetic flux density, and $\boldsymbol{J}_e$ is the current applied in the adjacent N layer. We recall that the corresponding transverse spin current at the interface creating the torque is $J_s^0 = \theta_{\mathrm{SHE}} \eta J_e$.

For a current applied along $\hat{\boldsymbol{x}}$, it can be shown from equation (86) that the deviation of the trajectory from the electron flow is: $\frac{v_y}{v_x} = -\frac{Q}{\alpha D}$ [W. Jiang *et al.*, *Nat. Phys.* **13**, 162 (2017)].

Summary

In this chapter, we have seen that *sd*-coupling between an incoming moment density $\boldsymbol{m}$ (equivalently spin density $\boldsymbol{s} = -\boldsymbol{m}/|\gamma|$) and a magnetization $\boldsymbol{M}$ results in a transfer of angular momentum. This angular momentum transfer alters spin accumulation and gives rise to a transverse spin transfer torque STT, which is decomposed in a damping-like component along $\boldsymbol{M} \times (\boldsymbol{m} \times \boldsymbol{M})$ and a field-like component along $\boldsymbol{m} \times \boldsymbol{M}$. The process of angular momentum transfer is governed by a complex number called spin mixing conductance $G^{\uparrow\downarrow}$, which relates to spin rotation and dephasing upon reflection and transmission, leading to a mixing of the eigenstates. It is most often used per unit area A per quantum conductance per spin channel $G_0/2$: $g^{\uparrow\downarrow} = \frac{G^{\uparrow\downarrow}}{AG_0/2}$, and its real and imaginary parts relate to the damping-like and field-like components, respectively. The transfer of angular momentum alters spin accumulation, through the addition of a relaxation channel, and magnetization dynamics, through the addition of a STT term. With a sufficiently large current, determined by the material's properties and system geometry, STT can trigger magnetization dynamics, precession or switching, of a uniform magnetization, and propagation of magnetic textures. Importantly, many effects in spintronics, like CPP-GMR, STT, spin pumping, and SMR, involve a transfer of angular momentum of the same type. They can thus be combined to accurately determine key spintronics parameters, like the spin mixing conductance $g^{\uparrow\downarrow}$ and the spin diffusion length l^*_{sf}. Other types of torques, called spin-orbit torques SOT, can influence transport and magnetization. These torques originate from spin-orbit coupling mechanisms. They are most often found in N/F bilayers for a current in-plane CIP configuration. In SOT mechanisms, not only a flux of angular momentum like in the STT mechanism but also non-equilibrium spin densities can be generated.

Chapter 5

Berry Curvature, Parity and Time Symmetries – Intrinsic AHE, SHE, QHE

In this chapter, we discuss an alternative approach to the previous chapters, which relied on extrinsic scattering. Here, we take advantage of the intrinsic symmetries of crystal and spin structures, represented by the different symmetry groups. Symmetries are particularly important because they have an impact on physical properties – and thus on spintronics – whether they are electrical (Hall effect family), thermal (Nernst effect family), or optical (Kerr effect family) [K. Brading and E. Castellani, *Symmetries in Physics: Philosophical Reflections*, Cambridge University Press, Cambridge (2003)]. This causal link can be formulated in the language of Berry formalism and is related to the breaking of spatial or temporal inversion symmetry. One of the objectives of this chapter is to demystify the Berry formalism (§5.1) which is now commonly used in condensed matter physics. With the example of the various Hall effects, whether unquantized (§5.2) or quantized (§5.3), we will present what the Berry formalism implies for intrinsic physical effects. The importance of breaking spatial or temporal inversion symmetries will be explained and illustrated by the example of several dispersion spectra commonly encountered in spintronics (§5.4). In a final section (§5.5), additional information will be provided on thermal analogs of the Hall effects.

5.1 Berry Curvature, Berry Phase

The Berry formalism stems from the adiabatic theorem in quantum mechanics, which states that a physical system remains in its instantaneous eigenstate, up to a phase, throughout the process of cyclic evolution, if a given perturbation is acting on it slowly enough.

The electronic transport in a material subjected to an external potential gradient is in fact determined from the Schrödinger equation $\mathcal{H}|\psi_n(\boldsymbol{k})\rangle = \varepsilon_n(\boldsymbol{k})|\psi_n(\boldsymbol{k})\rangle$, where the subscript indicates the nth band. This equation returns not one but two essential pieces of information to describe the dynamics of electrons:

DOI: 10.1051/978-2-7598-2917-0.c005

- the **energy band dispersion**, directly from the energy eigenvalues $\varepsilon_n(\boldsymbol{k}(t))$
- the **Berry curvature dispersion** $\boldsymbol{\Omega}_n(\boldsymbol{k}(t)) = \nabla_{\boldsymbol{k}} \times \langle \psi_n(\boldsymbol{k}) | i\nabla_{\boldsymbol{k}} | \psi_n(\boldsymbol{k}) \rangle$, from the Bloch wavefunction eigenstates $|\psi_n(\boldsymbol{k}(t))\rangle$ to within one phase $e^{i\gamma_n(\boldsymbol{k})}$. From this definition, we can already infer that the Berry curvature is large wherever the rate of change of spin in k-space is large, for example, close to the Dirac point of a Dirac cone [D. Xiao *et al.*, *Rev. Mod. Phys.* **82**, 1959 (2010)].

In the above expression, $\boldsymbol{k}(t)$ is the gauge invariant crystal momentum. It includes the time-varying external potential in the electron frame. It stems from the following transformation explained in detail in the review article by D. Xiao *et al.*, *Rev. Mod. Phys.* **82**, 1959 (2010): "A uniform $\boldsymbol{E}$ means that the electric potential $\phi(\boldsymbol{r})$ varies linearly in space and breaks the translational symmetry of the crystal so that Bloch's theorem cannot be applied. To avoid this difficulty, the electric field can enter through a uniform vector potential $\boldsymbol{A}(t)$ that changes in time."

We first give a set of definitions in k-space, followed by an example of a classical equivalent for a better understanding of this type of physics.

It is important to understand that the Berry curvature $\boldsymbol{\Omega}_n(\boldsymbol{k})$ is a local quantity defined at all points in k-space,

$$\boldsymbol{\Omega}_n(\boldsymbol{k}) = \nabla_{\boldsymbol{k}} \times \langle \psi_n(\boldsymbol{k}) | i\nabla_{\boldsymbol{k}} | \psi_n(\boldsymbol{k}) \rangle \tag{87}$$

It is a manifestation of the geometric properties of ψ_n in k-space. It is related to the Berry phase, which is a global quantity defined as the flux of $\boldsymbol{\Omega}_n(\boldsymbol{k})$ through a surface in k-space

$$\gamma_n(\boldsymbol{k}) = \oint \mathcal{A}_n(k)dk = \iint \Omega_n(\boldsymbol{k})d^2k \tag{88}$$

where $\mathcal{A}_n(\boldsymbol{k}) = \langle \psi_n(\boldsymbol{k}) | i\nabla_{\boldsymbol{k}} | \psi_n(\boldsymbol{k}) \rangle$ is the Berry connection or Berry potential. The Berry curvature arises from a cyclic adiabatic evolution, such as electron dynamics in a periodic crystal structure.

In specific cases, the Berry phase involves the integration of the Berry curvature over a closed surface. In these specific cases, mathematics tells us that the Berry phase is a multiple of 2π, leading to quantum effects like those we will describe in §5.3. The Chern number is then defined as

$$C_n = \frac{1}{2\pi} \oiint \Omega_n(\boldsymbol{k})d^2k = \text{integer} \tag{89}$$

An equivalent of the Berry curvature, potential, and phase in real(r)-space are the magnetic flux density $\boldsymbol{B}(\boldsymbol{r})$, vector potential $\boldsymbol{A}(\boldsymbol{r})$ and phase $\varphi(\boldsymbol{r}) = \oint A(\boldsymbol{r})dr = \int B(\boldsymbol{r})d^2r$ in the Aharonov–Bohm effect when electrons flow around a ring enclosing the magnetic flux density $\boldsymbol{B}(\boldsymbol{r})$ [C. Cohen-Tannoudji *et al.*, *Quantum Mechanics, Volume 1*, Wiley (1997)]. In this case, the equivalent of the Chern number is the Dirac monopole $q_m = \oiint B(\boldsymbol{r})d^2r = 2\Phi_0 \mathrm{i}$, where $\Phi_0 = \frac{h}{2e}$ is the magnetic flux quantum and i is an integer.

The way equations (87)–(89) are derived is detailed in D. Xiao *et al.*, *Rev. Mod. Phys.* **82**, 1959 (2010). We admit them here and start the reasoning from there.

We note that, in electron dynamics, the Berry phase relates to variations of $\boldsymbol{k}$ and is only dependent on the trajectory of this parameter. The Berry phase actually belongs to the broader class of geometric phases, grouping classical and quantum physical effects. It addresses the following question: what trajectory did the system follow?

An example of a classical geometric phase is presented below, to better understand what a geometric phase is.

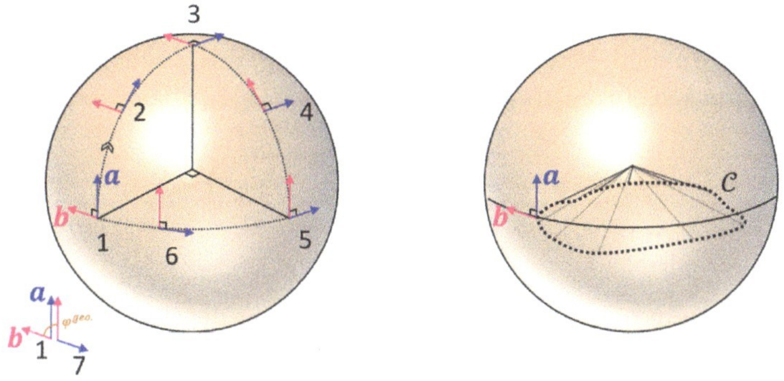

FIG. 51 – (Left) Representation of classical parallel transport on a closed contour on a sphere, showing how a frame spanned by two vectors $\boldsymbol{a}, \boldsymbol{b}$ can gain a geometric phase, here $\varphi^{\text{geo}} = \frac{\pi}{2}$, which depends only on the trajectory followed. (Right) General closed contour.

A frame $(\boldsymbol{a}, \boldsymbol{b})$ experiencing parallel transport on a closed contour $\mathcal{C}$ (figure 51, left) can gain a geometric phase φ^{geo}, which depends only on the trajectory followed. φ^{geo} corresponds to the solid angle subtented by $\mathcal{C}$. For the trajectory in figure 51, left, we observe that $\varphi^{\text{geo}} = \frac{\pi}{2}$. The solid angle subtended by the contour is indeed $\frac{1}{8}4\pi = \frac{\pi}{2}$.

For a general formalism for parallel transport on a closed contour $\mathcal{C}$ (figure 51, right), it is possible to define an initial state $(\boldsymbol{a}_i, \boldsymbol{b}_i)$ and a final one $(\boldsymbol{a}_f, \boldsymbol{b}_f)$ for the frame $(\boldsymbol{a}, \boldsymbol{b})$, linked by

$$\begin{aligned} \boldsymbol{a}_f &= \boldsymbol{a}_i \cos\varphi^{\text{geo}} - \boldsymbol{b}_i \sin\varphi^{\text{geo}} \\ \boldsymbol{b}_f &= \boldsymbol{a}_i \sin\varphi^{\text{geo}} + \boldsymbol{b}_i \cos\varphi^{\text{geo}} \end{aligned} \tag{90}$$

with φ^{geo} the solid angle subtended by the closed contour $\mathcal{C}$. If we consider the following initial state for the system

$$\boldsymbol{\psi}_i = \boldsymbol{a}_i + i\boldsymbol{b}_i \tag{91}$$

then the final state is

$$\boldsymbol{\psi}_f = \boldsymbol{a}_f + i\boldsymbol{b}_f = e^{i\varphi^{\text{geo}}} \boldsymbol{\psi}_i \tag{92}$$

Further information on geometric phases can be found, for example, in J. Dalibard's lecture, www.college-de-france.fr.

5.2 Unquantized Hall Effects – Intrinsic AHE, SHE

"It is now well recognized that information on the Berry curvature is essential in a proper description of the dynamics of Bloch electrons, which has various effects on transport and thermodynamic properties of crystals." [quoted from D. Xiao *et al.*, *Rev. Mod. Phys.* **82**, 1959 (2010)]. In this section, we will detail why Berry formalism has played a key role in understanding intrinsic effects in spintronics, such as the intrinsic Hall effects. Extrinsic contributions will be briefly presented at the end.

The Hall effects are manifested by the creation of a transverse current by applying a longitudinal potential gradient. Taking this statement into account, Ohm's law [equation (5)] reads

$$\boldsymbol{J}_e = \overline{\overline{\sigma}} \cdot \boldsymbol{E} = \begin{pmatrix} \sigma_{xx} & \sigma_{xy} & \sigma_{xz} \\ -\sigma_{xy} & \sigma_{yy} & \sigma_{yz} \\ -\sigma_{xz} & -\sigma_{yz} & \sigma_{zz} \end{pmatrix} \cdot \boldsymbol{E} \tag{93}$$

where σ_{ij} with $i \neq j$ are the transverse conductivity coefficients accounting for the Hall effects. In the most general case, the Hall current density $\boldsymbol{J}_H$ takes the form

$$\boldsymbol{J}_H = \boldsymbol{h} \times \boldsymbol{E} \tag{94}$$

with $\boldsymbol{h} = \left(\sigma_{zy}, \sigma_{xz}, \sigma_{yx}\right)^{\mathrm{T}}$ the Hall vector. We will consider that $\sigma_{xz} = \sigma_{yz} = 0$ in the next parts, for simplicity.

Obtaining a non-zero Hall effect is subject to the breaking of a set of spatial and temporal reversal symmetries ($\mathcal{PT}$, with $\mathcal{P}$ the parity or space inversion symmetry and $\mathcal{T}$ the time reversal symmetry) [N. Nagaosa *et al.*, *Rev. Mod. Phys.* **82**, 1539 (2010)]. These asymmetries can be achieved by several means, such as the application of external stimuli, coupling with an arrangement of magnetic moments, and being intrinsically linked to the magnetic point- and spin-symmetry groups. The Hall effect is then described by the following transverse resistivity, made of several contributions that will be defined in the following paragraphs

$$\rho_{xy} = R^N \mu_0 H + R^A \mu_0 M + \rho_{xy}^S + \rho_{xy}^T \tag{95}$$

■ Ordinary or normal Hall effect (OHE), $\rho_{xy}^N = R^N \mu_0 H$.

The OHE (figure 52, left) occurs in the presence of a perpendicular magnetic flux density $\boldsymbol{B} = \mu_0 \boldsymbol{H}$, when the trajectory of the electrons is deviated by the Lorentz force $-e\dot{\boldsymbol{r}} \times \boldsymbol{B}$. More generally, the change of momentum reads

$$\hbar \dot{\boldsymbol{k}} = -e\boldsymbol{E} - e\dot{\boldsymbol{r}} \times \boldsymbol{B}(\boldsymbol{r}) \tag{96}$$

with $\boldsymbol{k}$ the wavevector. In this case, the application of a magnetic field or flux density ensures $\mathcal{T}$-breaking, and the Lorentz force connects electrons to $\mathcal{T}$-breaking.

Note that the acronym OHE for ordinary Hall effect is also used to refer to the orbital Hall effect which is of a completely different nature [D. Go *et al.*, *Phys. Rev.*

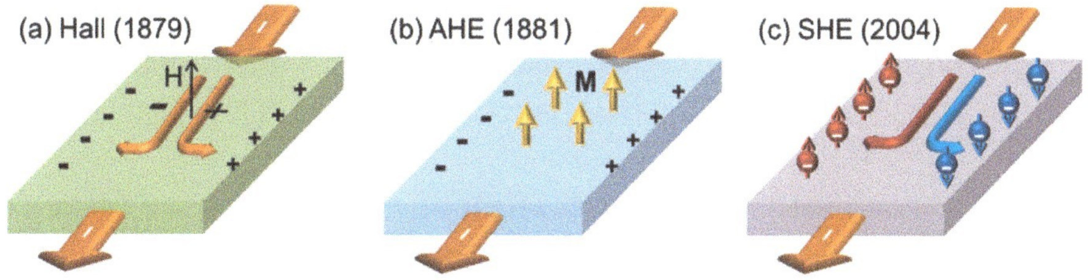

FIG. 52 – Representation of three types of Hall effects converting a longitudinal flow into a transverse one: (Left) the ordinary Hall effect (OHE), (Middle) the anomalous (also called extraordinary) Hall effect (AHE), and (Right) the spin Hall effect (SHE). In the last case, a net charge current is converted into a transverse spin current, but there is no net transverse charge current (see also §2.2). Reprinted with permission from IoP Science: C.-Z. Chang *et al.*, *J. Phys.: Condens. Matter* **28**, 123002 (2016). Copyright 2016.

Res. **2**, 033401 (2020)]. We also recall that an *overdot* indicates a time-derivative: $\dot{\boldsymbol{r}} = \partial \boldsymbol{r}/\partial t$.

In addition to the OHE, when an electron travels through specific spin structures (described by the magnetic point- and spin-symmetry groups) and spin textures (like skyrmions, §4.7) these can exert on it an effective or fictitious magnetic field or flux density and an associated effective Lorentz force. It thus produces the anomalous (or extraordinary) Hall effect $\rho_{xy}^{A} = R^{A}\mu_0 M$, the spontaneous anomalous Hall effect ρ_{xy}^{S}, and the topological Hall effect ρ_{xy}^{T}, depending on the nature of the spin structure and texture, and the interactions considered.

■ **Extraordinary or anomalous Hall effect (AHE),** $\rho_{xy}^{A} = R^{A}\mu_0 M$.

The AHE (figure 52, middle) describes the behavior of ferromagnetic materials with a non-zero net magnetization $\boldsymbol{M}$. A companion effect to the AHE is known as the **spin Hall effect (SHE)**, in which a transverse spin current is generated under the application of a longitudinal electric current (figure 52, right), whether the material is magnetic or not. In certain cases, the AHE can be considered as being the SHE for a spin-polarized material, and its magnitude equals the SHE magnitude times the polarization. An accurate description of the intrinsic contribution to AHE and SHE calls for a sophisticated model.

In the most general case, the effect can be linked to the spin-dependent electronic energy band structure, and formulated in the language of Berry curvature, as a geometric effect in reciprocal space. More generally, the change of position of an electron is

$$\boldsymbol{v}_n(\boldsymbol{k}) = \frac{\partial \varepsilon_n(\boldsymbol{k})}{\hbar \partial \boldsymbol{k}} - \dot{\boldsymbol{k}} \times \boldsymbol{\Omega}_n(\boldsymbol{k}) \tag{97}$$

with $\varepsilon_n(\boldsymbol{k})$ and $\boldsymbol{\Omega}_n(\boldsymbol{k})$ the energy- and Berry curvature-dispersion of the nth band. How equation (97) is obtained is extensively described in the review article by D. Xiao *et al.*, *Rev. Mod. Phys.* **82**, 1959 (2010). The first term of equation (97) relating velocity to the band dispersion is the one usually encountered. The second term results from internal $\mathcal{P}$- or $\mathcal{T}$-symmetry breaking of the crystal,[1] creating a non-zero Berry curvature $\boldsymbol{\Omega}_n(\boldsymbol{k})$, and leading to an effective Lorentz force and a related transverse velocity $-\dot{\boldsymbol{k}} \times \boldsymbol{\Omega}_n(\boldsymbol{k})$ after coupling the spin and lattice *via* a spin-orbit perturbation.

■ **Spontaneous anomalous Hall effect (SAHE), ρ^S_{xy}.**

Intrinsically, we do not need spin-orbit coupling to obtain a non-zero value for $\boldsymbol{\Omega}_n(\boldsymbol{k})$. The SAHE is specific to certain crystals belonging to spin-symmetry groups that allow internal $\mathcal{T}$-breaking despite zero net magnetization $M = 0$ (hence the name spontaneous). It is, for example, associated with a newly demonstrated unconventional *d*-wave magnetism, in analogy to *d*-wave superconductivity. This type of magnetism manifests itself as localized peaks of $\boldsymbol{\Omega}_n(\boldsymbol{k})$ and the associated electron dynamics is also described by equation (97). Those *d*-wave magnets open broad opportunities for designing and exploring unconventional quantum magnetism without geometrically frustrated lattices (see THE below) and with common elements (large spin-orbit interactions are not needed) [H. Reichlova *et al.*, arXiv:2012.15651 (2020) and L. Smejkal *et al.*, *Nat. Rev. Mater.* **7**, 482 (2022), see also figure 61].

■ **Topological Hall effect (THE), ρ^T_{xy}.**

The THE is specific to certain spin/moment $(\widehat{\boldsymbol{M}})$ chirality, which produces an effective magnetic field of the type $\widehat{\boldsymbol{M}} \cdot \left(\partial_x \widehat{\boldsymbol{M}} \times \partial_y \widehat{\boldsymbol{M}}\right)$. Those spin chiralities are, for example, encountered in specific antiferromagnets and in skyrmion textures. The phenomenon can also be formulated in the language of Berry curvature $\boldsymbol{\Omega}_n$ [equation (97)] as the result of a geometric effect in real space, as opposed to the geometric phase in k-space discussed in the previous paragraphs. Note that, for antiferromagnetic materials, the term topological used for the effect is related to the topology of the spin structure but does not imply a correspondence with a topological invariant, unlike the case of skyrmions (see §4.7 for a description of the companion skyrmion Hall effect). The THE is out of the scope of this book. The interested

---

[1]Non-zero $\boldsymbol{\Omega}_n$ requires breaking of $\mathcal{P}$- or $\mathcal{T}$-symmetry for the reason detailed below.

Under $\mathcal{P}$-symmetry, we have $\boldsymbol{k} \to -\boldsymbol{k}; \dot{\boldsymbol{k}} \to -\dot{\boldsymbol{k}}$; $\boldsymbol{v}_n(-\boldsymbol{k}) = -\boldsymbol{v}_n(\boldsymbol{k}); \varepsilon_n(-\boldsymbol{k}) = \varepsilon_n(\boldsymbol{k})$; $\frac{\partial \varepsilon_n(\boldsymbol{k})}{\hbar \partial \boldsymbol{k}} \to -\frac{\partial \varepsilon_n(\boldsymbol{k})}{\hbar \partial \boldsymbol{k}}$. Combining the above transformations with equation (97) imposes that $\boldsymbol{\Omega}_n(-\boldsymbol{k}) = \boldsymbol{\Omega}_n(\boldsymbol{k})$.

Under $\mathcal{T}$-symmetry, we have $\boldsymbol{k} \to -\boldsymbol{k}; \dot{\boldsymbol{k}} \to \dot{\boldsymbol{k}}$; $\boldsymbol{v}_n(-\boldsymbol{k}) = -\boldsymbol{v}_n(\boldsymbol{k}); \varepsilon_n(-\boldsymbol{k}) = \varepsilon_n(\boldsymbol{k})$; $\frac{\partial \varepsilon_n(\boldsymbol{k})}{\hbar \partial \boldsymbol{k}} \to -\frac{\partial \varepsilon_n(\boldsymbol{k})}{\hbar \partial \boldsymbol{k}}$. Combining the above transformations with equation (97) imposes that $\boldsymbol{\Omega}_n(-\boldsymbol{k}) = -\boldsymbol{\Omega}_n(\boldsymbol{k})$.

Hence, for a $\mathcal{PT}$ symmetric crystal, *i.e.*, for a crystal not breaking either $\mathcal{P}$- or $\mathcal{T}$-symmetry, $\boldsymbol{\Omega}_n(\boldsymbol{k}) = -\boldsymbol{\Omega}_n(\boldsymbol{k}) = 0$.

We note that more subtle symmetry considerations need to be taken into account, as some operations, like translation can restore symmetry, resulting in $\boldsymbol{\Omega}_n(\boldsymbol{k}) = 0$ despite initial $\mathcal{P}$- or $\mathcal{T}$-breaking (figure 61).

reader can find more information in *Topology in Magnetism*, J. Zang *et al.* (eds), Springer Nature (2018).

We next show how the contribution of Berry curvature to $\boldsymbol{v}_n(\boldsymbol{k})$ [equation (97)] connects with the Hall transverse conductivity σ_{xy} [equations (93) and (94)]. From Boltzmann's semiclassical transport theory [C. Kittel, *Introduction to Solid State Physics*, John Wiley & Sons (2004)], the charge current is

$$\boldsymbol{J}_e = -e\sum_n \int_{BZ} v_n(\boldsymbol{k})g(\varepsilon_n(\boldsymbol{k}))\frac{d^dk}{(2\pi)^d} \tag{98}$$

where d is the dimension of the system, BZ is the Brillouin zone (the integral is over the d dimensions) and $g(\varepsilon_n(\boldsymbol{k})) = f(\varepsilon_n(\boldsymbol{k})) + \delta g(\varepsilon_n(\boldsymbol{k}))$ the distribution function. We remind that this function accounts for the number of electrons with momentum $\boldsymbol{k}$ in the infinitesimal element d^dk whose position $\boldsymbol{r}$ is located in the infinitesimal element d^dr. $f(\varepsilon_n(\boldsymbol{k}))$ is the Fermi–Dirac distribution at equilibrium, and $\delta g(\varepsilon_n(\boldsymbol{k}))$ accounts for the deviation of the distribution from its equilibrium.

Combining equations (97) and (98), with $\dot{\boldsymbol{k}} = -\frac{e}{\hbar}\boldsymbol{E}$, it is possible to show that the total charge current takes the form

$$\begin{aligned}\boldsymbol{J}_e = &-\frac{e}{h}\sum_n \frac{1}{(2\pi)^{d-1}}\int_{BZ} \frac{\partial \varepsilon_n(\boldsymbol{k})}{\partial \boldsymbol{k}}\delta g(\varepsilon_n(\boldsymbol{k}))\,d^dk \\ &+\frac{e^2}{h}\boldsymbol{E}\times\sum_n \frac{1}{(2\pi)^{d-1}}\int_{BZ} \boldsymbol{\Omega}_n(\boldsymbol{k})f(\varepsilon_n(\boldsymbol{k}))\,d^dk\end{aligned} \tag{99}$$

The first term in equation (99) is the common one resulting directly from the energy dispersion. The second term represents the intrinsic contribution to the transverse conductivity. From $\boldsymbol{J}_e = \overline{\overline{\sigma}} \cdot \boldsymbol{E}$ this contribution is expressed as

$$\sigma_{xy} = \frac{e^2}{h}\sum_n \frac{1}{(2\pi)^{d-1}}\int_{BZ} \Omega_n(\boldsymbol{k})f(\varepsilon_n(\boldsymbol{k}))\,d^dk \tag{100}$$

The above equation describes the unquantized version of the Hall effect [see also equations (101) and (102)].

We note that, per definition, for a three-dimensional system, when $d = 3$ in equation (100), σ_{xy} is in S.m^{-1}, whereas, for a two-dimensional system, as in §5.3, when $d = 2$, σ_{xy} is in S, unit of conductance.

It is clear at this point that the electronic fingerprint of a material must contain two parts: the dispersion of the energy bands $\varepsilon_n(\boldsymbol{k})$ and that of the Berry curvature or total Berry curvature $\boldsymbol{\Omega}(\boldsymbol{k}) = \sum_n f(\varepsilon_n(\boldsymbol{k}))\boldsymbol{\Omega}_n(\boldsymbol{k})$, as shown for bcc Fe in figure 53. The peaks and valleys in the distribution of the total Berry curvature are due to pairs of unequally populated spin-orbit coupled bands, in a small k-interval, *i.e.*, wherever the rate of change of spin in k-space is large.

Fcc Pt is another prototypical material that illustrates the importance of the energy band topology and thus the need to provide the Berry curvature fingerprint. From figure 54, we observe that the intrinsic spin Hall conductivity σ_{xy}^z peaks just at the Fermi level, at $\varepsilon - \varepsilon_F = 0$. This makes Pt a material of choice for charge current

σ_{xy} $(S.cm^{-1})$	bcc Fe	fcc Ni	hcp Co
Fermi loop	750	−2275	478
Fermi loop (first term)	7	0	−4
Berry curvature	753	−2203	477
Previous theory	751[a]	−2073[b]	492[b]
Expt.	1032[c]	−646[d]	480[e]

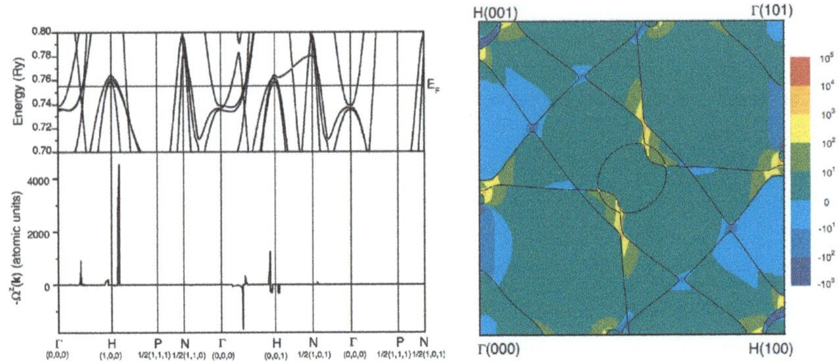

FIG. 53 – (Top) Table of the anomalous Hall conductivity σ_{xy}, computed in several ways, for bcc Fe, fcc Ni, and hcp Co. The third row shows the results obtained from calculations using the theoretical framework of Berry curvature. The other approximations are beyond the scope of the present book. The last row shows experimental data. The disagreement between theory and experiment for fcc Ni is due to the moderate onsite electron-electron correlations, which must be taken into account with more precision. Reprinted with permission from the American Physical Society: X. Wang *et al.*, *Phys. Rev. B* **76**, 195109 (2007). Copyright 2007. (Bottom Left) Electronic band structure and z-component of the total Berry curvature $\Omega^z(\boldsymbol{k}) = \sum_n f(\varepsilon_n(\boldsymbol{k}))\Omega^z_n(\boldsymbol{k})$ along symmetry lines of the reciprocal space, calculated for bulk Fe. (Bottom Right) Corresponding Fermi surface in the (010) plane, represented by solid lines, and Berry curvature in atomic units, represented by the color map. Atomic units mean $\propto (\pi/a)^{-2}$ with a the lattice parameter, where the proportionality factor depends on the number of k-points computed. For the calculations, the spin-orbit strength was 5.1 mRy (1 Ry $\sim$ 13.6 eV), and spin-orbit interactions were thus treated in a perturbative manner. Reprinted with permission from the American Physical Society: Y. Yao *et al.*, *Phys. Rev. Lett.* **92**, 037204 (2004). Copyright 2004.

to spin current interconversion (figure 52, right, see also §4.4 and exercise 6.6). The large value of σ^z_{xy} is attributed to the double degeneracies at the L and X points of high-symmetry, near the Fermi level, which are present even without spin-orbit coupling. From figure 54, it can be seen that when ε_F is lowered from its initial position, σ^z_{xy} reduces, then changes sign and peaks again at about −4.5 eV.

Let us now crudely consider that the Berry curvature fingerprint of the Pt material holds for all materials in the *5d* row (Ta, W, … , Au). The change from Au, Pt to W, Ta reduces the band filling and is equivalent to a lowering of ε_F. This rough approximation helps understand the dependence of σ^z_{xy} on band filling, and in particular, the change in sign between $\sigma^z_{xy}(\mathrm{Au, Pt})$ and $\sigma^z_{xy}(\mathrm{Ta, W})$ (figure 55).

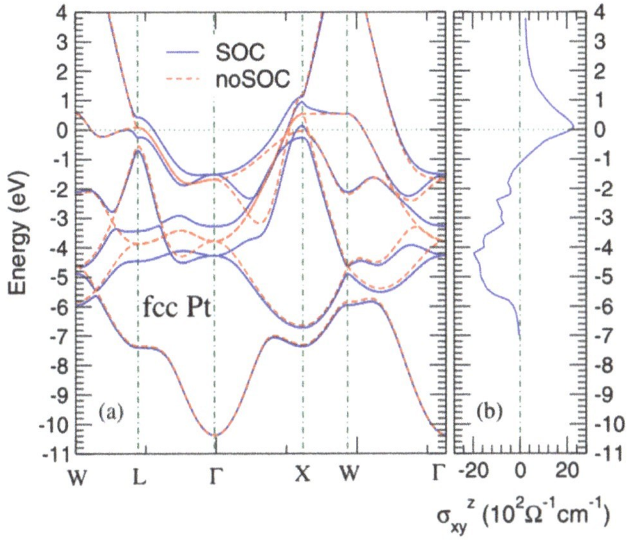

FIG. 54 – (Left) Electronic band structure along symmetry lines of the reciprocal space, calculated for bulk fcc Pt, when spin-orbit coupling (SOC) is turned off and on. (Right) Spin Hall conductivity (σ_{xy}^z) calculated for the same material, using the theoretical framework of the Berry formalism. Reprinted with permission from the American Physical Society: G. Y. Guo *et al.*, *Phys. Rev. Lett.* **100**, 096401 (2008). Copyright 2008.

■ Note about extrinsic contributions

The first term of equation (99) accounts for the longitudinal current. It may also contain transverse components in the presence of skew (or Mott) and side jump scattering on defects (figure 56), which impacts the non-equilibrium part $\delta g(\varepsilon_n(\boldsymbol{k}))$ of the distribution function $g(\varepsilon_n(\boldsymbol{k}))$ [N. Nagaosa *et al.*, *Rev. Mod. Phys.* **82**, 1539 (2010)]. Simply put, because the skew and side jump scattering mechanisms involve a one-step and a two-steps scattering process, respectively, their contributions to the total Hall resistivity ρ_{xy}^{total} are linearly and quadratically dependent on the longitudinal resistivity, respectively: $\rho_{xy}^{\text{skew}} \propto \rho_{xx}$ and $\rho_{xy}^{\text{side-jump}} \propto \rho_{xx}^2$. In addition, because the intrinsic Berry curvature contribution is independent of scattering, we have $\sigma_{xy}^{\text{intrinsic}} = \text{constant}$, and therefore $\rho_{xy}^{\text{intrinsic}} \propto \rho_{xx}^2$ from $\sigma_{xy}^{\text{intrinsic}} = \rho_{xy}^{\text{intrinsic}}/\rho_{xx}^2$. We note that the relation between $\sigma_{xy}^{\text{intrinsic}}$ and $\rho_{xy}^{\text{intrinsic}}$ is simply obtained from inversion of the resistivity matrix[2] considering that $\rho_{xx} \gg \rho_{xy}^{\text{intrinsic}}$, *i.e.*, when the determinant reduces to ρ_{xx}^2. From a simple scaling relation, the total transverse resistivity reads $\rho_{xy}^{\text{total}} = a\rho_{xx} + b\rho_{xx}^2$, where the coefficient a contains the skew scattering contribution and the coefficient b the side-jump scattering and intrinsic contributions.

---

[2] $$\begin{pmatrix} \sigma_{xx} & \sigma_{xy} \\ -\sigma_{xy} & \sigma_{xx} \end{pmatrix} = \begin{pmatrix} \rho_{xx} & \rho_{xy} \\ -\rho_{xy} & \rho_{xx} \end{pmatrix}^{-1} = \frac{1}{\rho_{xx}^2 + \rho_{xy}^2} \begin{pmatrix} \rho_{xx} & -\rho_{xy} \\ \rho_{xy} & \rho_{xx} \end{pmatrix}$$

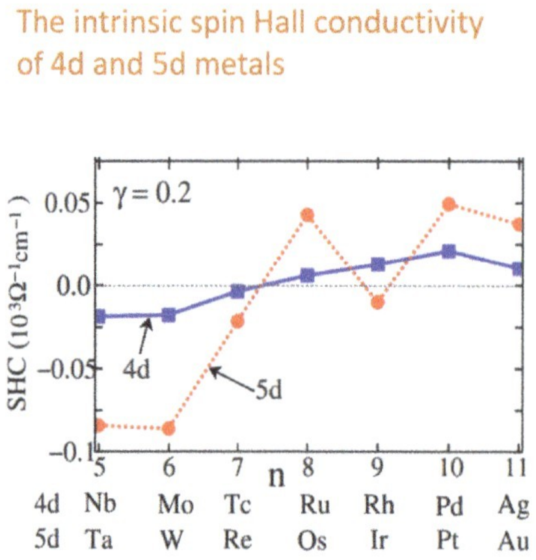

Metals	Structure	Electron number	SOI (Ry)
Nb	bcc	5 ($4d^4 5s^1$)	0.006
Mo	bcc	6 ($4d^5 5s^1$)	0.007
Tc	hcp	7 ($4d^6 5s^1$)	0.009
Ru	hcp	8 ($4d^7 5s^1$)	0.01
Rh	fcc	9 ($4d^8 5s^1$)	0.011
Pd	fcc	10 ($4d^{10} 5s^0$)	0.013
Ag	fcc	11 ($4d^{10} 5s^1$)	0.019
Ta	bcc	5 ($5d^3 6s^2$)	0.023
W	bcc	6 ($5d^4 6s^2$)	0.027
Re	hcp	7 ($5d^5 6s^2$)	0.025
Os	hcp	8 ($5d^6 6s^2$)	0.025
Ir	fcc	9 ($5d^9 6s^0$)	0.025
Pt	fcc	10 ($5d^9 6s^1$)	0.03
Au	fcc	11 ($5d^{10} 6s^1$)	0.03

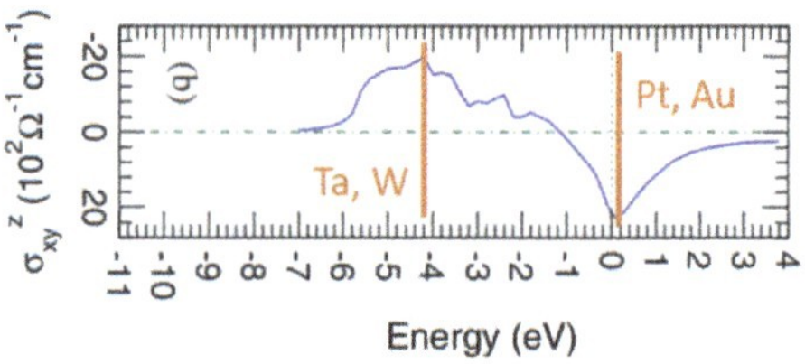

FIG. 55 – (Left) Spin Hall conductivity (SHC) at the Fermi level *versus* electron number (n), calculated for the *4d* and *5d* transition metals, using the theoretical framework of Berry curvature; (Right Top) table listing several parameters, for the *4d* and *5d* transition metals. SOI stands for spin-orbit interaction (1 Ry $\sim$ 13.6 eV). Reprinted with permission from the American Physical Society: T. Tanaka *et al.*, *Phys. Rev. B* **77**, 165117 (2008). Copyright 2008. (Right Bottom) Spin Hall conductivity (σ_{xy}^z) calculated for bulk fcc Pt, using the theoretical framework of Berry curvature. The vertical lines indicate the spin Hall conductivity at the Fermi level by considering that the calculation of σ_{xy}^z *vs. energy* in Pt stands for Ta, W, and Au, and that band filling moves the Fermi level. Adapted with permission from the American Physical Society: G. Y. Guo *et al.*, *Phys. Rev. Lett.* **100**, 096401 (2008). Copyright 2008.

Experimentally, a and b are obtained by varying the contribution from impurity scattering *via* changing the impurity content or temperature. The exact form of the scaling relation is actually more complex. It naturally involves residual (impurity-related) resistivity $\rho_{xx,T=0}$. The fact that the multiple scattering mechanisms compete with each other must be considered [D. Hou *et al.*, *Phys. Rev. Lett.* **114**, 217203 (2015)].

The mathematical description of extrinsic mechanisms is tedious. The models that describe these mechanisms go beyond the simplified scope of this book and are described in detail in, for example, the review article of N. Nagaosa *et al.*, *Rev. Mod. Phys.* **82**, 1539 (2010).

We note that the spin Hall angle is defined as $\theta_{\text{SHE}} = \rho_{xy}/\rho_{xx}$. Considering the intrinsic mechanism only, we thus obtain $\theta_{\text{SHE}} = \sigma_{xy}^{\text{intrinsic}}\rho_{xx}$ from $\sigma_{xy}^{\text{intrinsic}} = \rho_{xy}^{\text{intrinsic}}/\rho_{xx}^2$. In this case, the Hall angle is directly proportional to the longitudinal resistivity. This explains why the Hall angle for Au is much smaller than

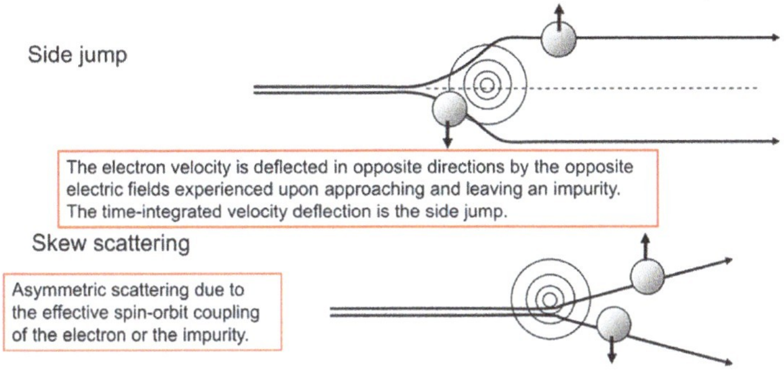

FIG. 56 – Schematics of two extrinsic mechanisms for the spin Hall effect: side jump and skew scattering. Adapted with permission from the American Physical Society: N. Nagaosa *et al.*, *Rev. Mod. Phys.* **82**, 1539 (2010). Copyright 2010.

for Pt, because Au is one order of magnitude less resistive than Pt, although the intrinsic transverse spin conductivity of Au and Pt are similar (figure 55). The factor of merit for spin-charge interconversion is therefore $\theta_{\mathrm{SHE}} l_{sf}^*$, because $l_{sf}^* \propto 1/\rho_{xx}$.

5.3 Quantum Hall Effect – QHE

In the previous section, we have seen the most general case of the unquantized version of the Hall effect. Quantization comes into play for certain types of energy band structures, like in some 2D electron gas and band insulators under large magnetic fields, especially when the energy levels or the bands become discrete Landau levels due to quantization of the cyclotron orbits of the particles. How the quantum mechanics of particles moving in a magnetic field connects to Landau levels with a finite density of states is, for example, demonstrated in D. Tong, arXiv:1606.06687. In such conditions, it is also possible to fold the Brillouin zone (BZ) to a torus [J. E. Moore and L. Balents, *Phys. Rev. B* **75**, 121306(R) (2007)], and projection or contraction onto a closed sphere under symmetry operations shows that points in the northern hemisphere are conjugate under $\mathcal{T}$ to those in the southern hemisphere. Therefore, the Berry phase or the integration of the Berry curvature on the closed BZ [equation (89)] becomes a multiple of 2π. Equation (100) simplifies as

$$\sigma_{xy} = \frac{e^2}{h} \sum_n \frac{1}{2\pi} \oiint_{BZ} \Omega_n(\boldsymbol{k}) d^2 k \tag{101}$$

Combining this last equation and equation (89) leads to

$$\sigma_{xy} = \frac{e^2}{h} \sum_n C_n \tag{102}$$

where C_n is the Chern number.

We recall that, per definition [equation (100)], for a two-dimensional system, as here, σ_{xy} is in S, unit of conductance.

The Hall effect then becomes quantized and is called the quantum Hall effect (QHE) [H. Weng *et al.*, *A-APPS Bulletin* **23**, 3 (2013)]. The quantum Hall conductivity can be considered as a counter of the number of fully filled Landau levels.

The QHE was discovered by K. von Klitzing *et al.* [*Phys. Rev. Lett.* **45**, 494 (1980)] in silicon-based 2D electron gas MOSFET (metal–oxide–semiconductor field-effect transistor) stacks. Under a strong magnetic field, the 2D electron gas bands are quantized into discrete (yet with finite width) Landau levels, which move up in energy when increasing the field. As a result, the number of filled Laudau levels is reduced with increasing field, every time a Landau level crosses the Fermi level. Consequently σ_{xy} (ρ_{xy}) reduces (increases) with field in a series of discrete jumps (figure 57).

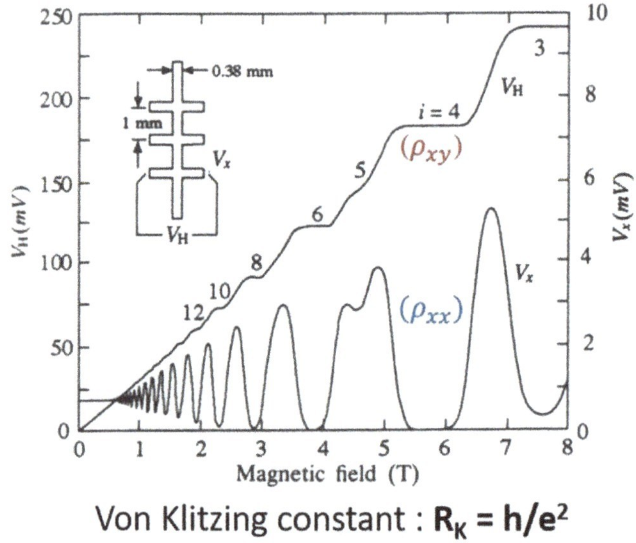

FIG. 57 – Quantum Hall effect demonstrated through quantized steps in the magnetic B-field-dependence of the transverse Hall voltage V_H. Each step corresponds to an incremental change in transverse conductivity: $\sigma_{xy} = \frac{1}{R_K}$i, where i is an integer [equation (102)]. Reprinted with permission from Springer Nature: K. von Klitzing *et al.*, *Nat. Rev. Phys.* **2**, 397 (2020). Copyright 2020.

The resistance $R_K = \frac{h}{e^2}$ was subsequently named the von Klitzing constant. It depends only on physical constants and is more stable and reproducible than any other resistance: $R_K = 25812.807\ \Omega$. For that reason, the QHE was essential for the redefinition of the kilogram in the international systems of units in 2019 (figure 58). Nowadays, SiC-substrate/graphene stacks are used as the QHE standard. The von Klitzing constant R_K relates to the conductance quantum G_0 as follows: $G_0 = 2\frac{e^2}{h} = \frac{2}{R_K}$. As an aside, we note that the Josephson effect [B. D. Josephson, *Phys. Lett.* **1**, 251 (1962)] in alternating current allows measurement of a similar quantity: $K_J = \frac{2e}{h}$, named the Josephson constant.

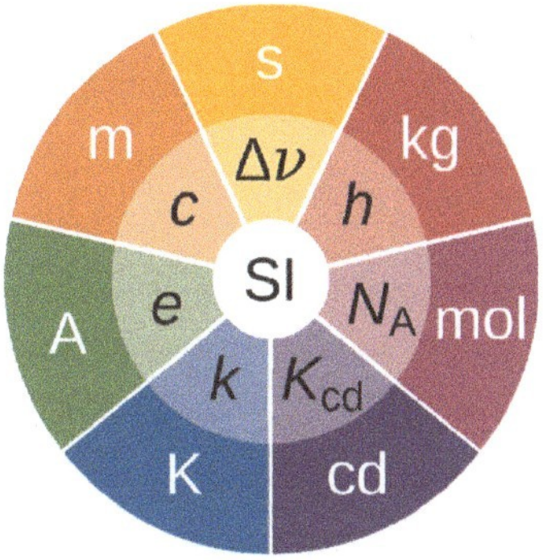

FIG. 58 – Diagrams of the base units of the old (outside) and new (inside) international systems. The new one, since 2019, is based on physical constants. From BIPM (bureau international des poids et mesures), A concise summary of the International System of Units, SI, https://www.bipm.org/en/home – CC BY License.

It is important to note that the QHE is one of the main manifestations of topology in condensed matter physics. For its discovery, K. von Klitzing was awarded the Nobel Prize in Physics in 1985. He opened the way to many other important related discoveries, including other types of quantum Hall effect [H. Weng *et al.*, *A-APPS Bulletin* **23**, 3 (2013)] for which other Nobel prizes have been awarded, for: "the discovery of a new form of quantum fluid with fractionally charged excitations" in 1998 (Nobels H. Störmer, D. Tsui, and R. Laughlin), "groundbreaking experiments regarding the two-dimensional material graphene" in 2010 (Nobels A. Geim and K. Novoselov), and "theoretical discoveries of topological phase transitions and topological phases of matter" in 2016 (Nobels D. J. Thouless, F. D. M. Haldane, and J. M. Kosterlitz).

5.4 Parity and Time Reversal Symmetries – $\mathcal{P}$, $\mathcal{T}$

We have seen in §5.2 that obtaining a non-zero Berry curvature and subsequently non-zero intrinsic Hall effects require breaking of $\mathcal{P}$- or $\mathcal{T}$-symmetry. In figure 59, we give some examples of how $\mathcal{P}$- or $\mathcal{T}$-symmetries alter the spin-dependent energy band dispersions. We next develop how some of those symmetries connect with the Berry curvature.

■ **Kramers – degenerate**

For free electrons (figure 59, top left), the band structures for majority moment ($\uparrow$)–spin($\Downarrow$) electrons and minority moment($\downarrow$)–spin($\Uparrow$) electrons are degenerate, and no-Hall effect is expected. The corresponding Hamiltonian contains the kinetic energy for free electrons and reads

$$\mathcal{H} = \frac{\hbar^2 \boldsymbol{k}^2}{2m_e} \tag{103}$$

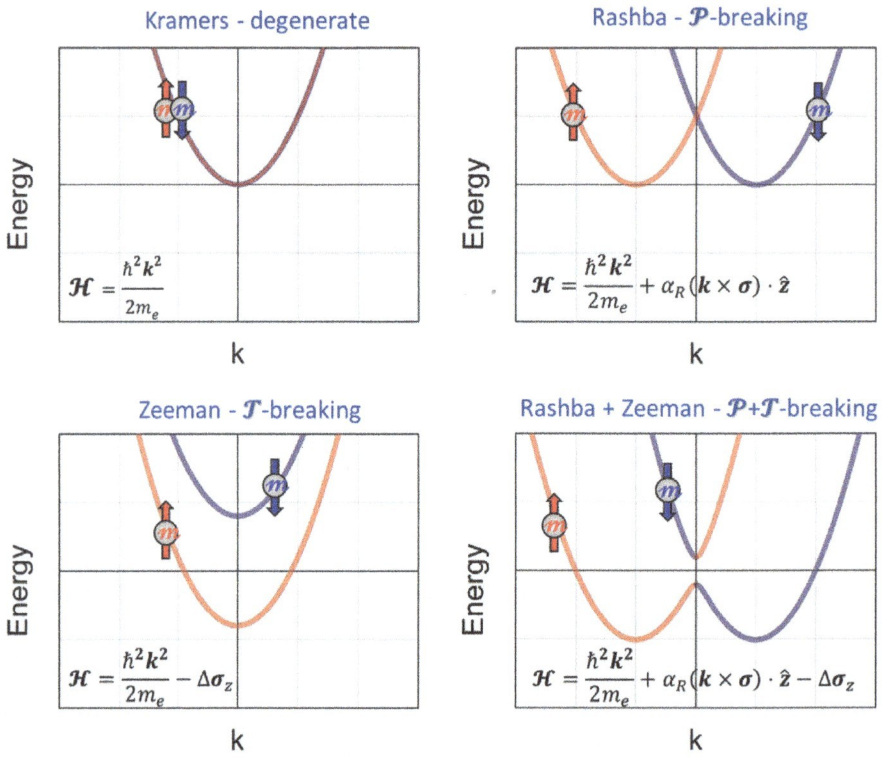

FIG. 59 – Band structures for majority moment(↑)–spin(⇓) electrons and minority moment (↓)–spin(⇑) electrons, corresponding to several parity and time symmetries, which are reflected in the Hamiltonian $\mathcal{H}$ involved.

■ Zeeman – $\mathcal{T}$-breaking

For a ferromagnet (figure 59, bottom left), the band structures for majority moment (↑)–spin(⇓) and minority moment(↓)–spin(⇑) electrons are split in energy (§2.3), leading to a non-zero magnetization and a Hall effect upon spin-orbit coupling. The corresponding Hamiltonian includes the Zeeman term

$$\mathcal{H} = \frac{\hbar^2 \boldsymbol{k}^2}{2m_e} - \Delta\boldsymbol{\sigma}_z \tag{104}$$

■ Rashba – $\mathcal{P}$-breaking

An example of Rashba parity breaking is when symmetry is extrinsically broken at an interface (figure 59, top right). The band structures for majority moment(↑)–spin (⇓) and minority moment(↓)–spin(⇑) electrons are thus shifted in momentum, leading to a Hall effect upon spin-orbit coupling. The spin angular momentum is then locked in a direction perpendicular to the linear momentum $\boldsymbol{k}$. The corresponding Hamiltonian includes the Rashba term (see also figure 49, left and middle and corresponding text)

$$\mathcal{H} = \frac{\hbar^2 \boldsymbol{k}^2}{2m_e} + \alpha_R(\boldsymbol{k} \times \boldsymbol{\sigma}) \cdot \hat{z} \tag{105}$$

Contributions triggered by spin-orbit coupling are called 'relativistic' because they involve a Lorentz transformation, $\mu_0 \boldsymbol{H}_{SO} = -\frac{(\boldsymbol{v} \times \boldsymbol{E})}{c^2}$. Conversely, the contributions triggered intrinsically with no need of spin-orbit coupling are called non-relativistic – although spin-orbit coupling is needed to couple spin and lattice in a final step. We note that the magnitude of relativistic contributions is limited by the ratio between electron velocity and speed of light, $\sim 10^{-4}$ in typical metals. These contributions thus require a large spin-orbit interaction to be significant. By contrast, non-relativistic contributions can be of the order of a few eV, at least one order of magnitude larger [*Topology in Magnetism*, J. Zang *et al.* (eds), Springer Nature (2018)].

■ Spin-polarized 2D electron gas with Rashba spin-orbit coupling – $\mathcal{P}$- and $\mathcal{T}$-breaking

A prototypical example illustrating Berry curvature physics is the case of a spin-polarized 2D electron gas with Rashba spin-orbit coupling (figure 59, bottom right) [S. Onoda *et al.*, *Phys. Rev. Lett.* **97**, 126602 (2006) and D. Xiao *et al.*, *Rev. Mod. Phys.* **82**, 1959 (2010)]. The combination of Rashba $\mathcal{P}$- and Zeeman $\mathcal{T}$-breaking results in a shift in the momentum of the band structures for majority moment($\uparrow$)–spin($\Downarrow$) electrons and minority moment($\downarrow$)–spin($\Uparrow$) electrons, accompanied by a gap opening at $k = 0$. The spin angular momentum is locked on the momentum $\boldsymbol{k}$. The Hamiltonian is

$$\mathcal{H} = \frac{\hbar^2 \boldsymbol{k}^2}{2m_e} + \alpha_R(\boldsymbol{k} \times \boldsymbol{\sigma}) \cdot \hat{z} - \Delta \boldsymbol{\sigma}_z \tag{106}$$

The associated eigenvalues (figure 60, top left) can be expressed as (exercise 6.10)

$$\varepsilon_{\pm}(k) = \frac{\hbar^2 k^2}{2m_e} \pm \sqrt{\alpha_R{}^2 k^2 + \Delta^2} = \varepsilon_0(k) \pm |\boldsymbol{b}(k)| \tag{107}$$

The corresponding eigenvectors are (exercise 6.10)

$$|\pm\rangle = \sqrt{1 \mp \frac{\Delta}{|\boldsymbol{b}(k)|}} |\psi_{\Uparrow}\rangle + ie^{i\phi} \sqrt{1 \pm \frac{\Delta}{|\boldsymbol{b}(k)|}} |\psi_{\Downarrow}\rangle \tag{108}$$

where ϕ is defined from $k_x + ik_y = ke^{i\phi}$, and $|\psi_{\Uparrow}\rangle = \frac{1}{\sqrt{2}} e^{i\boldsymbol{k}\cdot\boldsymbol{r}} \begin{pmatrix} 1 \\ 0 \end{pmatrix}$ and $|\psi_{\Downarrow}\rangle = \frac{1}{\sqrt{2}} e^{i\boldsymbol{k}\cdot\boldsymbol{r}} \begin{pmatrix} 0 \\ 1 \end{pmatrix}$ are the eigenvectors for free electrons.

From equations (87) and (108), it is then possible to show that the Berry curvature of the two subbands is (exercise 6.10)

$$\Omega_{\pm}^{z}(k) = \mp \frac{{\alpha_R}^2 \Delta}{2\left({\alpha_R}^2 k^2 + \Delta^2\right)^{3/2}} \tag{109}$$

The Berry curvature is non-zero around the gap, near $k = 0$ (figure 60, bottom left) if both Rashba ($\alpha_R \neq 0$) $\mathcal{P}$- and Zeeman ($\Delta \neq 0$) $\mathcal{T}$- breaking take place.

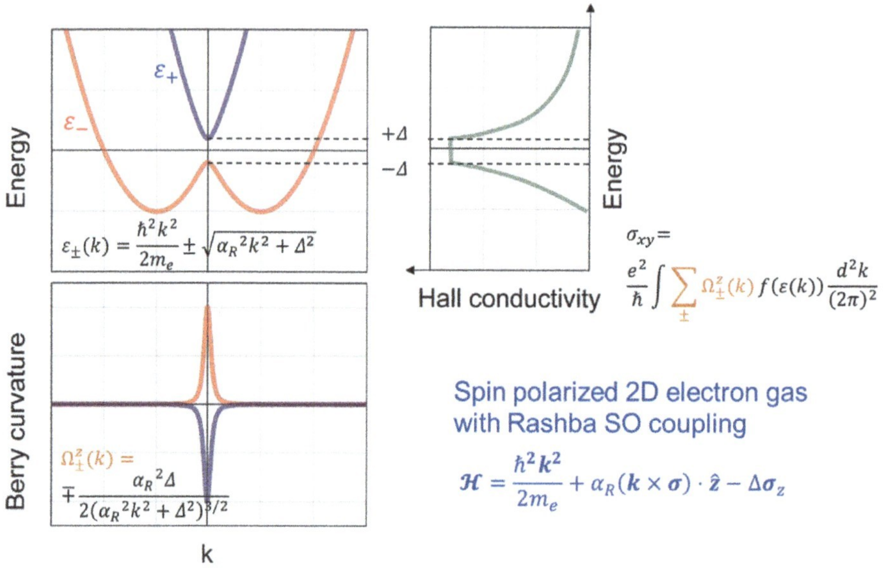

FIG. 60 – (Left) Band structure and Berry curvature for a spin-polarized 2D electron gas with Rashba spin-orbit (SO) coupling, described by the Hamiltonian ($\boldsymbol{\mathcal{H}}$). Note: beware the color code, when red/blue is commonly used to distinguish majority moment(↑)–spin(⇓) electrons and minority moment(↓)–spin(⇑) electrons bands (see figure 59, bottom right panel), it is also commonly used to distinguish the upper/lower band, no matter spin (like here). (Right) Corresponding Hall conductivity obtained from integration of the Berry curvature over the wavevector k for given energies.

The Hall conductivity σ_{xy} is further calculated numerically based on equation (100). From figure 60, right, we observe that for $\varepsilon \leq -\Delta$, σ_{xy} first increases when increasing the energy, as the contribution to the Berry curvature of the lower band $\Omega_{-}^{z}(k)$ is integrated. In the gap, for $-\Delta \leq \varepsilon \leq \Delta$, σ_{xy} saturates as the lower band has contributed fully. Finally, above the gap, for $\varepsilon \geq \Delta$, σ_{xy} reduces as both contributions $\Omega_{-}^{z}(k) + \Omega_{+}^{z}(k)$ are now integrated, and the contribution to the Berry curvature of the upper band opposes that of the lower band.

Causes	Effects		
Symmetry of the spin structure	Antiferromagnetism $n = m_1 - m_2$	Hall effect $h = (\sigma_{zy}, \sigma_{xz}, \sigma_{yx})$	Material
Two-sublattice (1 m_1 and 1 m_2 in the unit cell)	$t_{1/2}\mathcal{T}(n) = -(-n) = n$ Even Compatible	$\neq$ $t_{1/2}\mathcal{T}(h) = +(-h) = -h$ Odd Incompatible	
	$\mathcal{PT}(n) = -(-n) = n$ Even Compatible	$\neq$ $\mathcal{PT}(h) = +(-h) = -h$ Odd Incompatible	
	$t_{1/2}\mathcal{C}_{2y}(n) = n$ $t_{1/2}\mathcal{C}_{2x}\mathcal{T}(n) = -(-n) = n$ Even Compatible	$=$ $t_{1/2}\mathcal{C}_{2y}(h) = h$ Even – Compatible along y $\neq$ $t_{1/2}\mathcal{C}_{2x}\mathcal{T}(h) = -(-h) = -h$ Odd – Incompatible along x	RuO$_2$, CrSb, MnTe
Multi-sublattice (2 or 4 m_1 and 2 or 4 m_2 in the unit cell)	$t_{1/2}\mathcal{M}_y\mathcal{R}_S(n) = n$ Even Compatible	$=$ $t_{1/2}\mathcal{M}_y\mathcal{R}_S(h) = h$ Even Compatible	Epi-Mn$_5$Si$_3$ (hexagonal)
	$t_{1/2}\mathcal{M}_y\mathcal{R}_S\mathcal{T}(n) = n$ Even Compatible	$\neq$ $t_{1/2}\mathcal{M}_y\mathcal{R}_S\mathcal{T}(h) = -h$ Odd Incompatible	Bulk-Mn$_5$Si$_3$ (orthorhombic)

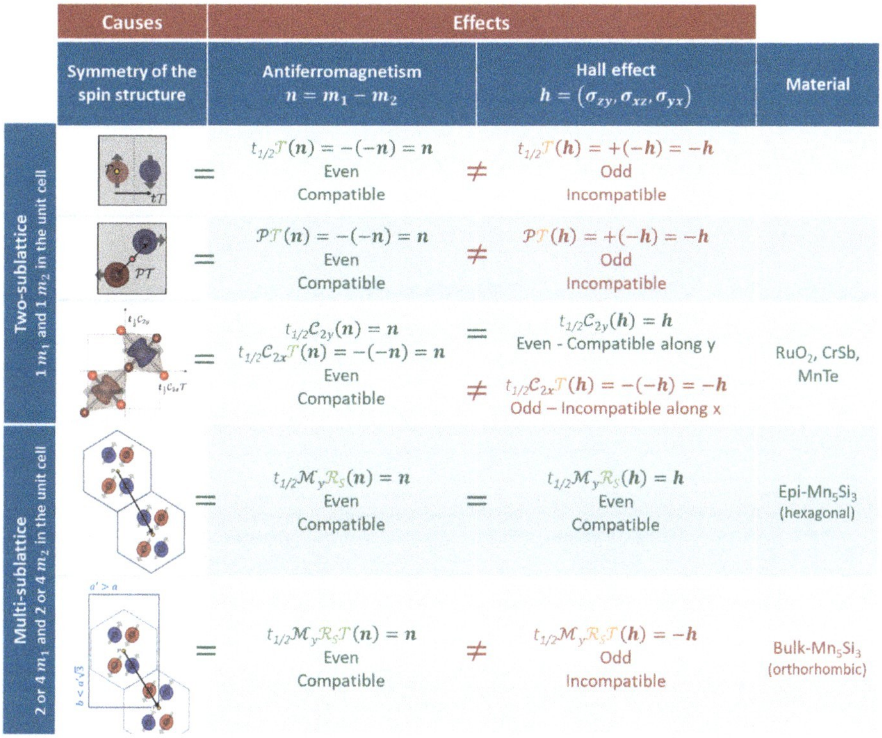

FIG. 61 – Illustration of a symmetry analysis to establish which of the antiferromagnetic structures shown on the left column may host the spontaneous anomalous Hall effect. Here, m_1 and m_2 are the sublattice magnetizations, normalized, and n is the Néel vector, which defines the magnetic order of an antiferromagnet. The $t_{1/2}$ half-unit cell translation, $\mathcal{C}_{2x(y)}$ 2-fold rotation-axis symmetries, $\mathcal{M}_y$ mirror symmetry and $\mathcal{R}_S$ spin rotation allow describing the antiferromagnetic structures. Applying the corresponding symmetry combinations to the Néel vector n is naturally compatible with the proposed structures. However, applying these symmetry combinations to the Hall vector h is compatible in some structures only, meaning that only these structures may host both antiferromagnetism and the spontaneous anomalous Hall effect. Note that $\mathcal{M}_y$ and $\mathcal{R}_S$ are unlocked, and description *via* spin groups is therefore relevant for the two corresponding cases (see also figure 62). The illustration is made after the information contained in L. Smejkal *et al.*, *Nat. Rev. Mater.* **7**, 482 (2022) and H. Reichlova *et al.*, arXiv:2012.15651.

To conclude this section, we note that the Curie–Neumann principle of symmetry states: "when certain causes produce certain effects, the elements of symmetry of the causes must be found in the effects produced". Based on a symmetry analysis, it is therefore possible – and now systematically done in practice – to establish which type of symmetry is compatible with one physical effect, such as the Hall effect, and which type of related crystal can *a priori* host the effect (figure 61). The reciprocal principle is not true as the effects produced are often more symmetrical than the causes.

We also remark that a thorough symmetry analysis calls for the use of group theory, whether the relevant group is the crystallographic point group (no magnetism), the magnetic point group (in which magnetic and crystal rotation operations are locked; it corresponds to the case of large spin-orbit coupling) or the spin group (in which magnetic and crystal rotation operations are partially independent of each other (figure 62) [H. Reichlova *et al.*, arXiv:2012.15651 (2020), L. Smejkal *et al.*, *Nat. Rev. Mater.* **7**, 482 (2022), P. Liu *et al.*, *Phys. Rev. X* **12**, 021016 (2022)].

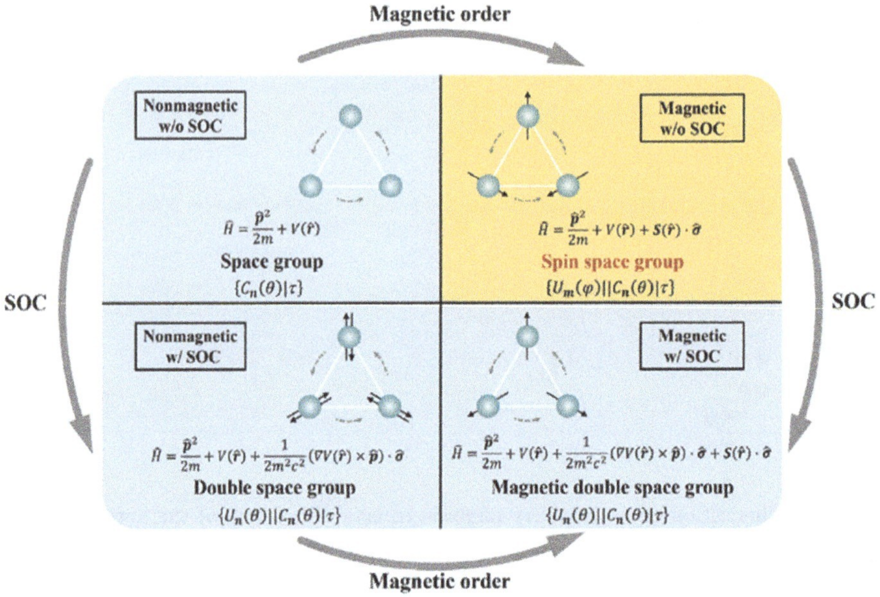

FIG. 62 – Illustration describing the symmetry of solids. $S(\widehat{\boldsymbol{r}})$ and $V(\widehat{\boldsymbol{r}})$ in the Hamiltonian $\widehat{H}$, stand for exchange field due to the distribution of local magnetic moments, and crystal potential energy, respectively. $\widehat{\sigma}$ is the Pauli operator associated with the electron spin and $\widehat{\boldsymbol{p}}$ the momentum operator associated with the linear momentum. Spatial and spin rotations are described by $C_n(\theta)$ and $U_{n(m)}(\theta, \phi)$, where n and m denote the rotation axes, and θ and ϕ are the rotation angles. The notations w and w/o SOC correspond to with and without spin-orbit coupling. w SOC (Bottom panels), the spatial and spin rotations are locked. w/o SOC (Top panels), the spin and spatial rotations are completely or partially decoupled. In nonmagnetic cases (Left panels) spatial rotation only is considered or an unconstrained spin rotation is added. In magnetic cases (Right panels), spin rotation is constrained by the magnetic order of the system. From P. Liu *et al.*, *Phys. Rev. X* **12**, 021016 (2022) – CC BY license.

5.5 Thermal Nernst Counterparts

Electrical and thermal transport are closely related. Therefore, there are thermal counterparts to electrical effects in spintronics. Analogously to Ohm's law

$\boldsymbol{J}_e = -\overline{\overline{\sigma}} \cdot \boldsymbol{\nabla}\phi$ with $\overline{\overline{\sigma}}$ the electric conductivity tensor, and ϕ the electric potential [equation (5)], which describes the magnitude of the induced electric current in response to a potential gradient, the magnitude of the induced thermoelectric current in response to a temperature gradient is as follows

$$\boldsymbol{J}_e = -\overline{\overline{\alpha}} \cdot \boldsymbol{\nabla} T \tag{110}$$

with $\overline{\overline{\alpha}}$ the thermoelectric conductivity tensor. This tensor takes into account, among others, the Seebeck effect in the diagonal terms, with $\boldsymbol{\alpha} = \boldsymbol{\sigma} \cdot \boldsymbol{S}$, where $\boldsymbol{S}$ are the Seebeck coefficients, and the Nernst effect in the off-diagonal terms.

Analogous to the Hall effect, the Nernst effect accounts for the creation of a transverse voltage in response to a longitudinal thermal gradient. The thermoelectric transverse resistivity ρ^{ϱ}_{xy} (with $\alpha_{xy} = \rho^{\varrho}_{xy}/{\rho^{\varrho}_{xx}}^2$) can be described similarly [equation (95)] by a sum of normal, anomalous, spontaneous anomalous and topological contributions as

$$\rho^{\varrho}_{xy} = R^{\varrho,N}\mu_0 H + R^{\varrho,A}\mu_0 M + \rho^{\varrho,S}_{xy} + \rho^{\varrho,T}_{xy} \tag{111}$$

In the most general case, the thermoelectric and electric coefficients are related by the Mott relation [equation (112)]. It has been shown that the validity of these relations extends to intrinsic contributions based on the Berry curvature formalism [D. Xiao *et al.*, *Phys. Rev. Lett.* **97**, 026603 (2006)]. Therefore, the intrinsic anomalous Nernst conductivity is also related to the intrinsic anomalous Hall conductivity, calculated in §5.2, *via* the following Mott relation

$$\alpha_{xy} = -2e\frac{g_0}{G_0}\frac{\partial \sigma_{xy}(\varepsilon_F)}{\partial \varepsilon} \tag{112}$$

where $G_0 = 2\frac{e^2}{h} = \frac{2}{R_K}$ is the electrical conductance quantum and $g_0 = \frac{\pi^2 k_B^2 T}{3h}$ is the thermal conductance quantum, to which quantized thermoelectric effects relate [J. Noky *et al.*, *J. Phys. Commun.* **5**, 045007 (2021)], in a manner analogous to the link between G_0 and quantized electric effects like the quantum Hall effect (§5.3).

From equation (112), we see that α_{xy} is related to the derivative of σ_{xy} with respect to energy, and thus to the derivative of the DOS with respect to energy, in a similar way to other thermoelectric properties. Therefore, α_{xy} will be maximized as the edges of the bands are approached (figure 60, from $\frac{\partial \sigma_{xy}}{\partial \varepsilon}$). At the microscopic scale, the application of a temperature gradient populates the higher energy states and depopulates the lower energy states, on the hot side. The system returns to equilibrium by diffusion. A net current is created if and only if the diffusion of the high-energy and low-energy states is asymmetric, hence the derivative in the Mott relation.

More generally, the links between thermoelectric and electrical coefficients are valid whether the contributions are intrinsic or extrinsic, and whether the effect involves diagonal or off-diagonal terms. For example, how the spin-dependent Seebeck effect alters the drift-diffusion equation for spin accumulation is shown in exercise 6.3. All this contributes to the richness of the cross-effects that exist (figure 63),

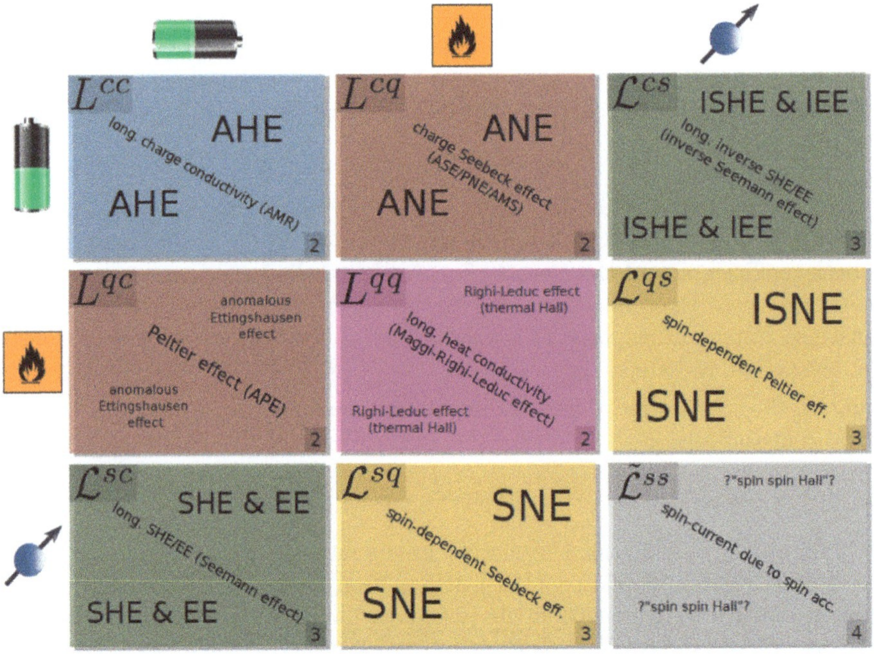

FIG. 63 – Illustration of the wide range of charge, spin, and heat-related effects and how these effects are intertwined. In this illustration, $\mathcal{L}^{ij}$ is a tensor that connects a charge ($i = c$), spin ($i = s$), or heat ($i = q$) current to a force of an electric $\nabla\phi$ ($j = c$), spin $\nabla\mu_s$ ($j = s$), or heat ∇T ($j = q$) nature. The number on the bottom right of each panel indicates the order of the corresponding tensor. In this book, we discussed the AHE, ANE, ISHE, and SHE. Courtesy of H. Ebert. See also M. Seeman *et al.*, *Phys. Rev. B* **92**, 155138 (2015) for a full description.

the corollary being that these effects can be difficult to distinguish from each other in experiments [S. R. Boona *et al.*, *Energy Environ. Sci.* **7**, 885 (2014); K.-i. Uchida *et al.*, *J. Phys. Soc. Jpn* **90**, 122001 (2021); M. Seeman *et al.*, *Phys. Rev. B* **92**, 155138 (2015)]. To complete this comment, in the same way that it is possible to pump spins and then act on the magnetization *via* the application of an electric current through a potential gradient, it is also possible to induce similar effects using a temperature gradient, and cross-check the results to confirm the findings.

Summary

In this chapter, we have seen that Berry formalism, which derives from the adiabatic theorem of quantum mechanics, is appropriate to describe the cyclic adiabatic evolution of Bloch electrons in a periodic crystal structure. In addition to the conventional energy band dispersion $\varepsilon_n(\boldsymbol{k})$, the Berry curvature dispersion $\Omega_n(\boldsymbol{k})$ is an essential ingredient for a correct description of electron dynamics. As a thumb rule, the Berry curvature is large, wherever the rate of change of spin in k-space is large. From its impact on the electron velocity in k-space $\boldsymbol{v}_n(\boldsymbol{k}) = \frac{\partial \varepsilon_n(\boldsymbol{k})}{\hbar \partial \boldsymbol{k}} - \dot{\boldsymbol{k}} \times \boldsymbol{\Omega}_n(\boldsymbol{k})$, the Berry curvature gives rise to intrinsic spintronics effects, like the AHE, the SHE, the THE, and Nernst thermal counterparts. It also leads to quantized physical effects, like the QHE, under specific dimensionality and cyclic conditions. $\mathcal{P}$- or $\mathcal{T}$-symmetry breaking is essential for the Berry curvature to be non-zero, therefore underlying the importance of considerations on symmetry groups. Finally, thermal and electric effects are related *via* the Mott relation $\alpha_{xy} = -2e\frac{g_0}{G_0}\frac{\partial \sigma_{xy}(\varepsilon_F)}{\partial \varepsilon}$, which can be used in spintronics.

Chapter 6

Exercises and Solutions

This chapter provides a series of ten comprehensive exercises with solutions. They are specifically designed to illustrate the ideas in the previous chapters and lead to other spintronic effects based on these ideas.

The exercises deal with:

- **Anisotropic magnetoresistance (AMR)** (chapter 2)
- **Domain wall anisotropic magnetoresistance (DWAMR)** (chapter 3)
- **Drift-diffusion equation for spin accumulation (μ_s)** (chapters 3–5)
- **Spin conductivity mismatch (CPP-GMR)** (chapter 3)
- **Intra-band scattering – intrinsic damping (α_0)** (chapters 2 and 4)
- **Spin pumping (SP) and inverse spin Hall effect (ISHE)** (chapters 4 and 5)
- **Spin pumping (SP) – additional damping (α_p)** (chapter 4)
- **Spin Hall magnetoresistance (SMR)** (chapters 2–4)
- **Harmonic analysis of the anomalous Hall voltage ($V_{xy}^{\omega}, V_{xy}^{2\omega}$)** (chapters 4 and 5)
- **Intrinsic anomalous Hall and Nernst effects (AHE, ANE)** (chapter 5)

We recall that a List of symbols and units, as well as the corresponding formulas, is provided on page 150. When there are several possible definitions in the literature, the definition used in this book is made explicit, and the differences in prefactor depending on the articles and books consulted are also made explicit.

6.1 Anisotropic Magnetoresistance (AMR) – chapter 2

Introduction/Reminder

The anisotropic magnetoresistance (AMR) effect (§2.6) refers to the dependence of the electrical resistivity ρ on the relative angle θ between the applied electrical

DOI: 10.1051/978-2-7598-2917-0.c006

current $\boldsymbol{I}$ and the magnetization $\boldsymbol{M}$ of a magnetic material [W. Thomson, *Proc. Roy. Soc.* **8**, 546 (1857); T. McGuire and R. Potter, *IEEE Trans.Magn.* **11**, 1018 (1975)]. It is a bulk property caused by anisotropic mixing of majority moment(↑)–spin(⇓) electrons and minority moment(↓)–spin(⇑) electrons conduction bands, induced by the spin-orbit interaction.

In the reference frame set by $\boldsymbol{M}$ (figure 64), the electric field $\boldsymbol{E}$ and the current density $\boldsymbol{J}_e$ are linked by the resistivity tensor $\bar{\bar{\rho}}$

$$\boldsymbol{E} = \bar{\bar{\rho}}\boldsymbol{J}_e$$
$$\begin{pmatrix} E_\parallel \\ E_\perp \end{pmatrix} = \begin{pmatrix} \rho_\parallel & 0 \\ 0 & \rho_\perp \end{pmatrix} \begin{pmatrix} J_\parallel \\ J_\perp \end{pmatrix} \tag{113}$$

with $\rho_\parallel$ and $\rho_\perp$, the resistivities for $\boldsymbol{J}_e \parallel \boldsymbol{M}$ and $\boldsymbol{J}_e \perp \boldsymbol{M}$, respectively (§2.6).

■ Questions

(a) Find the expressions of E_x and E_y *vs.* J_x, J_y, $\rho_\parallel$, $\rho_\perp$ and θ, in the experimental reference frame defined by $(\hat{\boldsymbol{x}}, \hat{\boldsymbol{y}})$, (figure 64).

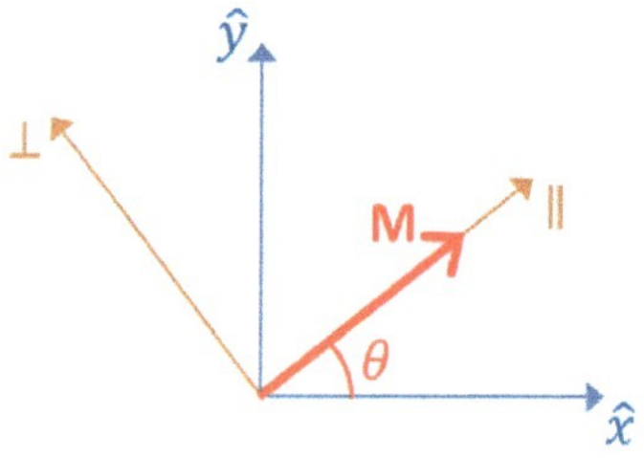

FIG. 64 – Illustration showing the two reference frames considered.

(b) In practice, how would you proceed experimentally to measure $\rho_\parallel$ and $\rho_\perp$? What is the order of magnitude of the values of field, current, voltage, etc. that you think you would use or measure?

(c) Experimental measurements of Ni films at room temperature returned $\rho_\parallel = 8.2$ $\mu\Omega$.cm and $\rho_\perp = 8$ $\mu\Omega$.cm. Calculate the AMR ratio $\frac{\Delta\rho}{\rho} = \frac{\rho_\parallel - \rho_\perp}{\rho_\perp}$ for Ni. Can you think of any practical use of the AMR effect?

(d) We now consider a 'Union Jack' device shown in figure 65, a magnetization pointing along $\hat{\boldsymbol{x}}$, a film of thickness d, and a time-dependent square wave current density $J_{ij}(t)$. Plot the time(t)-variation of: V_{15}, and V_{37} for J_{15}; V_{84}, and V_{26} for J_{84}; V_{73}, and V_{15} for J_{73}; V_{62}, and V_{84} for J_{62}.

(e) An antiferromagnet is used instead of a ferromagnet. This antiferromagnet has two collinear sublattices with magnetizations pointing towards opposite directions ($M_1 = -M_2$). The total magnetization is $M = M_1 + M_2 = 0$. Do you think that using AMR is appropriate to characterize this type of magnetic material?

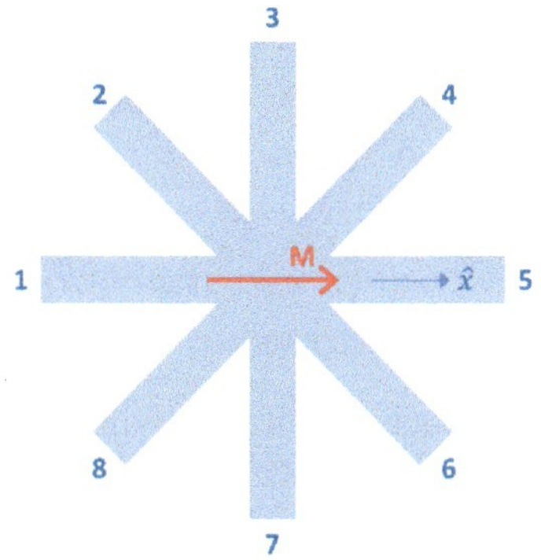

FIG. 65 – Illustration of the 'Union Jack' device considered.

Solutions

(a) A change of basis can be done by using the appropriate rotation matrix

$$\begin{pmatrix} E_\parallel \\ E_\perp \end{pmatrix} = \begin{pmatrix} \cos\theta & \sin\theta \\ -\sin\theta & \cos\theta \end{pmatrix} \begin{pmatrix} E_x \\ E_y \end{pmatrix} \text{ and } \begin{pmatrix} J_\parallel \\ J_\perp \end{pmatrix} = \begin{pmatrix} \cos\theta & \sin\theta \\ -\sin\theta & \cos\theta \end{pmatrix} \begin{pmatrix} J_x \\ J_y \end{pmatrix} \tag{114}$$

Combining equations (113) and (114) gives

$$\begin{pmatrix} \cos\theta & \sin\theta \\ -\sin\theta & \cos\theta \end{pmatrix} \begin{pmatrix} E_x \\ E_y \end{pmatrix} = \begin{pmatrix} \rho_\parallel & 0 \\ 0 & \rho_\perp \end{pmatrix} \begin{pmatrix} \cos\theta & \sin\theta \\ -\sin\theta & \cos\theta \end{pmatrix} \begin{pmatrix} J_x \\ J_y \end{pmatrix},$$

$$\begin{pmatrix} E_x \\ E_y \end{pmatrix} = \begin{pmatrix} \cos\theta & \sin\theta \\ -\sin\theta & \cos\theta \end{pmatrix}^{-1} \begin{pmatrix} \rho_\parallel & 0 \\ 0 & \rho_\perp \end{pmatrix} \begin{pmatrix} \cos\theta & \sin\theta \\ -\sin\theta & \cos\theta \end{pmatrix} \begin{pmatrix} J_x \\ J_y \end{pmatrix},$$

$$\begin{pmatrix} E_x \\ E_y \end{pmatrix} = \frac{1}{\cos^2\theta + \sin^2\theta} \begin{pmatrix} \cos\theta & -\sin\theta \\ \sin\theta & \cos\theta \end{pmatrix} \begin{pmatrix} \rho_\parallel & 0 \\ 0 & \rho_\perp \end{pmatrix} \begin{pmatrix} \cos\theta & \sin\theta \\ -\sin\theta & \cos\theta \end{pmatrix} \begin{pmatrix} J_x \\ J_y \end{pmatrix},$$

$$\begin{pmatrix} E_x \\ E_y \end{pmatrix} = \begin{pmatrix} \cos\theta\rho_\parallel & -\sin\theta\rho_\perp \\ \sin\theta\rho_\parallel & \cos\theta\rho_\perp \end{pmatrix} \begin{pmatrix} \cos\theta & \sin\theta \\ -\sin\theta & \cos\theta \end{pmatrix} \begin{pmatrix} J_x \\ J_y \end{pmatrix},$$

$$\begin{pmatrix} E_x \\ E_y \end{pmatrix} = \begin{pmatrix} \cos^2\theta\rho_\parallel + \sin^2\theta\rho_\perp & \cos\theta\sin\theta\left(\rho_\parallel - \rho_\perp\right) \\ \cos\theta\sin\theta\left(\rho_\parallel - \rho_\perp\right) & \sin^2\theta\rho_\parallel + \cos^2\theta\rho_\perp \end{pmatrix} \begin{pmatrix} J_x \\ J_y \end{pmatrix},$$

$$\begin{pmatrix} E_x \\ E_y \end{pmatrix} = \begin{pmatrix} \rho_\parallel - \left(\rho_\parallel - \rho_\perp\right)\sin^2\theta & \cos\theta\sin\theta\left(\rho_\parallel - \rho_\perp\right) \\ \cos\theta\sin\theta\left(\rho_\parallel - \rho_\perp\right) & \rho_\perp - \left(\rho_\perp - \rho_\parallel\right)\sin^2\theta \end{pmatrix} \begin{pmatrix} J_x \\ J_y \end{pmatrix} \tag{115}$$

For $J_y = 0$, we obtain

$$E_x = \left[\rho_{\|} - \left(\rho_{\|} - \rho_{\perp}\right)\sin^2\theta\right] J_x,$$

$$E_y = \left(\rho_{\|} - \rho_{\perp}\right)\cos\theta\sin\theta J_x.$$

Note that, in this case, a transversal voltage drop $\propto E_y$ is obtained when $\theta \neq 0$ and $\theta \neq \pi/2$. By analogy, this effect is called the planar Hall effect. It is noteworthy that the terminology planar Hall effect is misleading because the phenomenon at play is unrelated to the Hall effects. The terminology transverse AMR is sometimes preferred.

(b) Possible option 1: use 4-point resistance measurements and a known geometry [I. Miccoli *et al.*, *J. Phys. Cond. Mat.* **27**, 223201 (2015)].
Set $I_x \neq 0$ and $I_y = 0$.
Monitor V_x, to get R_x.
Use a magnet to apply an external magnetic field at $\theta = 0$ and get $\rho_{\|}$; and at $\theta = 90°$ and get $\rho_{\perp}$.
Possible option 2: use 4-point resistance measurements and a known geometry. Use a magnet to apply an external magnetic field and set $\theta = 0$. Set $I_x \neq 0$ and $I_y = 0$ and monitor V_x, to get R_x and then $\rho_{\|}$. Set $I_x = 0$ and $I_y \neq 0$ and monitor V_y, to get R_y and then $\rho_{\perp}$.
Orders of magnitude: mm device, mA, tenth of Ohms, mV.

(c) For Ni, $\frac{\Delta\rho}{\rho} \sim 2.5\%$ at room temperature.
The AMR effect was used in the first generation of sensors, like read heads in hard disk drives. Nowadays, most sensors are based on tunnel magnetoresistance with orbital filtering. Typical values of magnetoresistance are now greater than 100%. The temperature dependence of AMR sensors can still be a plus for some applications.

(d) From equation (115) we can deduce the following:
For J_{15}, we have $\theta = 0$

$$E_{15} = \rho_{\|} J_{15}$$

$$E_{37} = 0$$

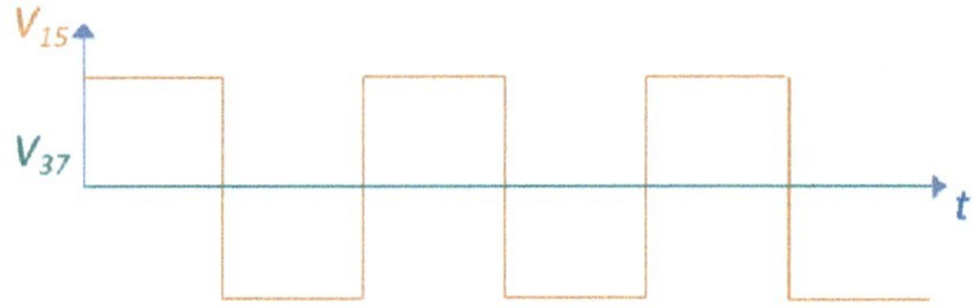

For J_{84}, we have $\theta = -45°$

$$E_{84} = \left(\rho_{\parallel} + \rho_{\perp}\right) J_{84}/2$$

$$E_{26} = -\left(\rho_{\parallel} - \rho_{\perp}\right) J_{84}/2$$

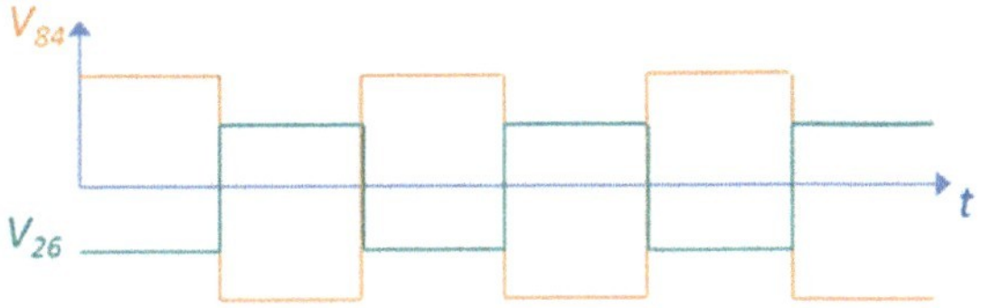

For J_{73}, we have $\theta = -90°$

$$E_{73} = \rho_{\perp} J_{73}/d$$

$$E_{15} = 0$$

For J_{62}, we have $\theta = -135°$

$$E_{62} = \left(\rho_{\parallel} + \rho_{\perp}\right) J_{84}/2$$

$$E_{84} = \left(\rho_{\parallel} - \rho_{\perp}\right) J_{62}/2$$

(e) The AMR effect is even in magnetization, *i.e.*, it is invariant on magnetization reversal. It can be shown that $\frac{\Delta\rho}{\rho} \propto (\boldsymbol{M}.\boldsymbol{J}_e)^2$. The AMR responses from the two sublatices add up. Using the AMR effect is hence appropriate for characterizing/detecting a collinear antiferromagnet [P. Wadley *et al.*, *Science* **351**, 587 (2016)].

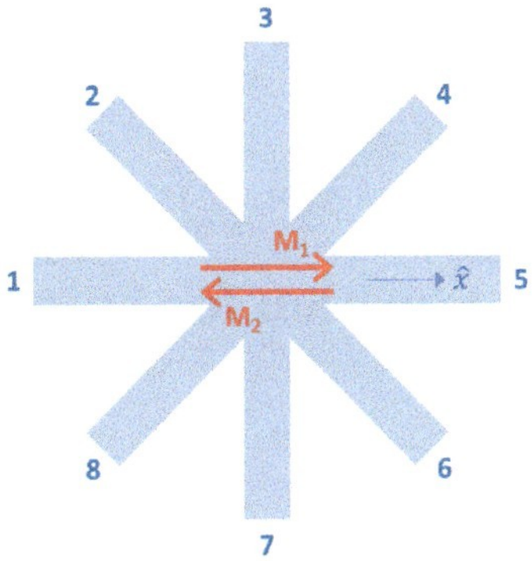

6.2 Domain Wall Anisotropic Magnetoresistance (DWAMR) – chapter 3

■ Introduction/Reminder

Several types of magnetic domain walls (DWs) in magnetic wires (§3.5) are depicted in figure 66. An electrical current (*I*) flows along the wires.

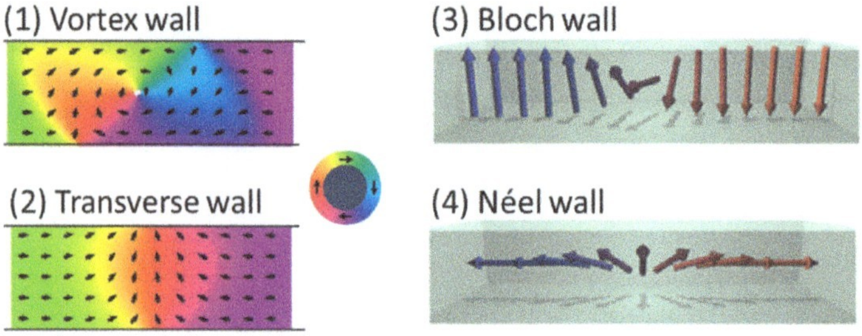

FIG. 66 – Representation of 4 types of magnetic domain walls. Adapted with permission from IoP Science: R. L. Stamps *et al.*, *J. Phys. D: Appl. Phys.* **47**, 333001 (2014). Copyright 2014.

■ Questions

(a) Which of these DWs display a signal of anisotropic magnetoresistance $\frac{\Delta\rho}{\rho}$ compared to the uniformly magnetized wire, and why? We recall that $\frac{\Delta\rho}{\rho} = \frac{\rho_{\text{with DW}} - \rho_{\text{without DW}}}{\rho_{\text{without DW}}} = \frac{\rho_{\text{DW}} - \rho_0}{\rho_0}$ (equation 43).

(b) What is the sign of the effect in each case?

■ Solutions

(a) Case (3), $I \perp M$ with and without DW => $\frac{\Delta\rho}{\rho} = 0$.

Cases (1), (2) and (4), $I \parallel M$ without DW and $I \perp M$ in the DW => $\frac{\Delta\rho}{\rho} \neq 0$.

(b) The high resistivity state is obtained when $I \parallel M$ => $\rho_0 > \rho_{\text{DW}}$.

Cases (1), (2) and (4) => $\frac{\Delta\rho}{\rho} < 0$.

6.3 Drift-Diffusion Equation for Spin Accumulation (μ_s) – chapters 3–5

■ Introduction/Reminder

The distinct partial densities of current at the interface between materials of different types create a spin accumulation (or chemical potential imbalance) $\mu_s = -(\mu^{\uparrow} - \mu^{\downarrow})$

(§3.2). Relaxation towards equilibrium conditions causes spins to diffuse near the interface, resulting in a diffusion-type equation for μ_s: $\nabla^2\mu_s - \frac{\mu_s}{l_{sf}^{*2}} = 0$ [equation (27)].

Questions

(a) Demonstrate equation (27), using the method proposed in §3.2.

(b) Deduce how the law of total angular momentum conservation $\nabla \cdot \boldsymbol{J}_s = -e\left(\frac{\delta n^{\uparrow}}{\tau_{sf}^{\uparrow}} - \frac{\delta n^{\downarrow}}{\tau_{sf}^{\downarrow}}\right)$ [equation (26)] takes the form $\nabla \cdot \boldsymbol{J}_s = \frac{\sigma^*(1-\beta^2)}{4el_{sf}^{*2}}\mu_s$ [first term in equation (64), in §4.3].

(c) Show that equation (27) becomes $\nabla^2\mu_s - \frac{\mu_s}{l_{sf}^{*2}} - e\left(\frac{dS_s}{dT}(\nabla T)^2 + S_s\nabla^2 T\right) = 0$, with $S_s = -\left(S^{\uparrow} - S^{\downarrow}\right)$, if one considers the spin-dependent Seebeck effect (§5.5), *i.e.* $\boldsymbol{j}^{\uparrow(\downarrow)} = \frac{\sigma^{\uparrow(\downarrow)}}{e}\nabla\mu^{\uparrow(\downarrow)} - \sigma^{\uparrow(\downarrow)}S^{\uparrow(\downarrow)}\nabla T$.

Solutions

(a) Step 1: generalized Ohm's law

$$\nabla\mu^{\uparrow(\downarrow)} = \frac{e}{\sigma^{\uparrow(\downarrow)}}\boldsymbol{j}^{\uparrow(\downarrow)}$$

$$\Rightarrow \nabla\left(\mu^{\uparrow} - \mu^{\downarrow}\right) = e\left(\frac{1}{\sigma^{\uparrow}}\boldsymbol{j}^{\uparrow} - \frac{1}{\sigma^{\downarrow}}\boldsymbol{j}^{\downarrow}\right)$$

$$\Rightarrow \nabla^2\left(\mu^{\uparrow} - \mu^{\downarrow}\right) = e\left(\frac{1}{\sigma^{\uparrow}}\nabla\cdot\boldsymbol{j}^{\uparrow} - \frac{1}{\sigma^{\downarrow}}\nabla\cdot\boldsymbol{j}^{\downarrow}\right)$$

Step 2: charge and spin current

$$\boldsymbol{J}_e = \boldsymbol{j}^{\uparrow} + \boldsymbol{j}^{\downarrow}$$

$$\boldsymbol{J}_s = -\left(\boldsymbol{j}^{\uparrow} - \boldsymbol{j}^{\downarrow}\right)$$

Step 3: charge conservation

$$\nabla\cdot\boldsymbol{J}_e = 0$$

$$\Rightarrow \nabla\cdot\boldsymbol{j}^{\uparrow} = -\nabla\cdot\boldsymbol{j}^{\downarrow}$$

$$\Rightarrow \nabla^2\left(\mu^{\uparrow}-\mu^{\downarrow}\right)=e2\nabla\cdot\boldsymbol{j}^{\uparrow}\left(\frac{1}{\sigma^{\uparrow}}+\frac{1}{\sigma^{\downarrow}}\right)$$

Step 4: total angular momentum conservation (only spin-flip is considered in this calculation)

$$\nabla\cdot\boldsymbol{J}_s=-e\left(\frac{\delta n^{\uparrow}}{\tau_{sf}^{\uparrow}}-\frac{\delta n^{\downarrow}}{\tau_{sf}^{\downarrow}}\right)$$

$$\Rightarrow -2\nabla\cdot\boldsymbol{j}^{\uparrow}=-e\left(\frac{\delta n^{\uparrow}}{\tau_{sf}^{\uparrow}}-\frac{\delta n^{\downarrow}}{\tau_{sf}^{\downarrow}}\right)$$

$$\Rightarrow \nabla^2\left(\mu^{\uparrow}-\mu^{\downarrow}\right)=e^2\left(\frac{1}{\sigma^{\uparrow}}+\frac{1}{\sigma^{\downarrow}}\right)\left(\frac{\delta n^{\uparrow}}{\tau_{sf}^{\uparrow}}-\frac{\delta n^{\downarrow}}{\tau_{sf}^{\downarrow}}\right) \tag{116}$$

Step 5: obtaining the diffusion equation for spin accumulation
Near equilibrium, we have

$$\mu^{\uparrow(\downarrow)}\propto\frac{\delta n^{\uparrow(\downarrow)}}{N^{\uparrow(\downarrow)}}$$

$$\Rightarrow \nabla^2\left(\mu^{\uparrow}-\mu^{\downarrow}\right)=e^2\left(\frac{1}{\sigma^{\uparrow}}+\frac{1}{\sigma^{\downarrow}}\right)\left(\frac{N^{\uparrow}\mu^{\uparrow}}{\tau_{sf}^{\uparrow}}-\frac{N^{\downarrow}\mu^{\downarrow}}{\tau_{sf}^{\downarrow}}\right)$$

At equilibrium we have $n^{\uparrow(\downarrow)}=N^{\uparrow(\downarrow)}$, and equilibrium recovery after stochastic reversal imposes

$$k=\frac{N^{\uparrow}}{\tau_{sf}^{\uparrow}}=\frac{N^{\downarrow}}{\tau_{sf}^{\downarrow}}$$

$$\Rightarrow \nabla^2\left(\mu^{\uparrow}-\mu^{\downarrow}\right)=e^2k\left(\frac{1}{\sigma^{\uparrow}}+\frac{1}{\sigma^{\downarrow}}\right)\left(\mu^{\uparrow}-\mu^{\downarrow}\right)$$

$$\Rightarrow \nabla^2\mu_s-\frac{\mu_s}{l_{sf}^{*\,2}}=0 \tag{117}$$

with $\frac{1}{l_{sf}^{*\,2}}=\frac{e^2k}{\sigma^{\uparrow}}+\frac{e^2k}{\sigma^{\downarrow}}=\frac{1}{l_{sf}^{\uparrow\,2}}+\frac{1}{l_{sf}^{\downarrow\,2}}$

(b) Combining equations (116) and (117) gives

$$\frac{\mu_s}{l_{sf}^{*\,2}} = -e^2\left(\frac{1}{\sigma^{\uparrow}} + \frac{1}{\sigma^{\downarrow}}\right)\left(\frac{\delta n^{\uparrow}}{\tau_{sf}^{\uparrow}} - \frac{\delta n^{\downarrow}}{\tau_{sf}^{\downarrow}}\right)$$

$$\Rightarrow \frac{\mu_s}{l_{sf}^{*\,2}} = -e^2\left(\frac{\sigma^{\uparrow} + \sigma^{\downarrow}}{\sigma^{\uparrow}\sigma^{\downarrow}}\right)\left(\frac{\delta n^{\uparrow}}{\tau_{sf}^{\uparrow}} - \frac{\delta n^{\downarrow}}{\tau_{sf}^{\downarrow}}\right)$$

Using $\sigma^{\uparrow(\downarrow)} = \frac{\sigma^*}{2}(1-(+)\beta)$ the above expression transforms in

$$\frac{\mu_s}{l_{sf}^{*\,2}} = -\frac{4e^2}{\sigma^*(1-\beta^2)}\left(\frac{\delta n^{\uparrow}}{\tau_{sf}^{\uparrow}} - \frac{\delta n^{\downarrow}}{\tau_{sf}^{\downarrow}}\right)$$

Further introducting equation (26), one gets

$$\boldsymbol{\nabla}\cdot\boldsymbol{J}_s = \frac{\sigma^*(1-\beta^2)}{4el_{sf}^{*\,2}}\mu_s$$

(c) Taking into account the spin-dependent Seebeck effect, the generalized Ohm's law writes:

$$\boldsymbol{\nabla}\mu^{\uparrow(\downarrow)} = \frac{e}{\sigma^{\uparrow(\downarrow)}}\boldsymbol{j}^{\uparrow(\downarrow)} + eS^{\uparrow(\downarrow)}\boldsymbol{\nabla}T$$

$$\Rightarrow -\boldsymbol{\nabla}(\mu_s) = e\left(\frac{1}{\sigma^{\uparrow}}\boldsymbol{j}^{\uparrow} - \frac{1}{\sigma^{\downarrow}}\boldsymbol{j}^{\downarrow}\right) - eS_s\boldsymbol{\nabla}T$$

$$\Rightarrow -\boldsymbol{\nabla}^2(\mu_s) = e\left(\frac{1}{\sigma^{\uparrow}}\boldsymbol{\nabla}\cdot\boldsymbol{j}^{\uparrow} - \frac{1}{\sigma^{\downarrow}}\boldsymbol{\nabla}\cdot\boldsymbol{j}^{\downarrow}\right) - e\boldsymbol{\nabla}(S_s\boldsymbol{\nabla}T)$$

$$\Rightarrow -\boldsymbol{\nabla}^2(\mu_s) = e\left(\frac{1}{\sigma^{\uparrow}}\boldsymbol{\nabla}\cdot\boldsymbol{j}^{\uparrow} - \frac{1}{\sigma^{\downarrow}}\boldsymbol{\nabla}\cdot\boldsymbol{j}^{\downarrow}\right) - e\boldsymbol{\nabla}(S_s)\boldsymbol{\nabla}T - eS_s\boldsymbol{\nabla}^2T$$

$$\Rightarrow -\boldsymbol{\nabla}^2(\mu_s) = e\left(\frac{1}{\sigma^{\uparrow}}\boldsymbol{\nabla}\cdot\boldsymbol{j}^{\uparrow} - \frac{1}{\sigma^{\downarrow}}\boldsymbol{\nabla}\cdot\boldsymbol{j}^{\downarrow}\right) - e\frac{dS_s}{dT}(\boldsymbol{\nabla}T)^2 - eS_s\boldsymbol{\nabla}^2T$$

Based on point (a) we obtain:

$$\Rightarrow \boldsymbol{\nabla}^2\mu_s - \frac{\mu_s}{l_{sf}^{*\,2}} - e\left(\frac{dS_s}{dT}(\boldsymbol{\nabla}T)^2 + S_s\boldsymbol{\nabla}^2T\right) = 0$$

A practical use of this type of thermoelectric effect and related equation is for example shown in A. Slachter *et al.*, *Nat. Phys.* **6**, 879 (2010).

6.4 Spin Conductivity Mismatch (CPP-GMR) – chapter 3

■ Introduction/Reminder

We consider a F/I/N/I/F multilayer (§3.3) consisting of two metallic ferromagnets (F) made of the same material, one metallic non-magnetic spacer (N) of thickness d_N and spin diffusion length $l^*_{sf,N}$, and two interfaces (I) (Inset of figure 67).

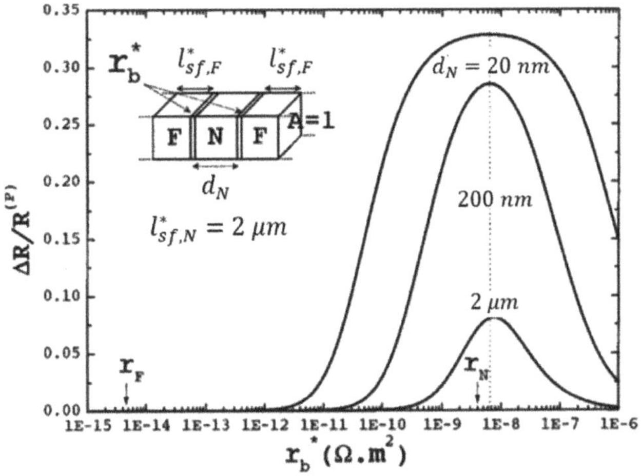

FIG. 67 – Calculated magnetoresistance ($\Delta R/R^{(P)}$) *vs.* interfacial (I) spin resistance area product $r_b^* = r_b(1-\gamma^2)$, for a F/I/N/I/F multilayer. Adapted with permission from the American Physical Society: A. Fert and H. Jaffrès, *Phys. Rev. B* **64**, 184420 (2001). Copyright 2001.

For $d_N \ll l^*_{sf,N}$, when an electric current is applied perpendicular to the sample plane, it can be shown that the difference ΔR between $R^{(AP)}$ and $R^{(P)}$ is

$$\Delta R = \frac{2(\beta r_F + \gamma r_b)^2}{(r_F + r_b) + \frac{r_N}{2}\left[1 + \left(\frac{r_b}{r_N}\right)^2\right]\frac{d_N}{l^*_{sf,N}}}$$

and that

$$R^{(P)} = 2(1-\beta^2)r_F + \frac{d_N}{l^*_{sf,N}} r_N + 2(1-\gamma^2)r_b + 2\frac{(\beta-\gamma)^2 r_F r_b + (\beta^2 r_F + \gamma^2 r_b)\frac{d_N}{2l^*_{sf,N}} r_N}{(r_F + r_b) + \frac{d_N}{2l^*_{sf,N}} r_N}$$

with

$R^{(AP)}$, the resistance of the stack when the magnetizations of the two F layers are antiparallel.

$R^{(P)}$, the resistance of the stack when the magnetizations of the two F layers are parallel.

r_N, the spin resistance area product of the N layer (resistivity ρ_N times $l^*_{sf,N}$).

r_F, the spin resistance area product of the F layers.

r_b, the spin resistance area product due to the interfaces, I ($r_b = r_{\text{interface}}$ in §3.4).

β, the spin polarization of the F layers.

γ, the spin polarization due to the interfaces, I

■ Questions

(a) Justify why experiments on current perpendicular to plane magnetoresistance usually fulfill the following condition: $d_N \ll l^*_{sf,N}$.

(b) Express $\Delta R/R^{(P)}$ (as a function of β only) for the case with no interfacial spin resistance ($r_b = 0$) and with only metals ($r_N \sim r_F$). Reminder: $d_N \ll l^*_{sf,N}$.

(c) Express $\Delta R/R^{(P)}$, still for the case with no interfacial spin resistance ($r_b = 0$) but with a semiconducting spacer ($r_N \gg r_F$). Consider $\frac{r_N}{r_F} \gg \frac{2l^*_{sf,N}}{d_N} \gg 1$.

(d) From (c), what can you conclude on spin injection in a semiconductor?

(e) Still for the case of a semiconducting spacer but with a non-zero interfacial spin resistance ($r_b \neq 0$), we admit that it is possible to calculate the maximum value of magnetoresistance as $\Delta R/R^{(P)} = \gamma^2/(1-\gamma^2)$.
- From this expression, what is the result of adding a large interfacial spin resistance on spin-injection in a semiconductor?
- What type of material that you know of gives a large spin-dependent resistance?

(f) Calculate the three values of $\Delta R/R^{(P)}$ from questions (b), (c) and (e), for: $\beta = 0.5$, $\gamma = 0.5$, $r_N = 4 \times 10^{-9}\,\Omega.\text{m}^2$ (for the case of the semiconducting spacer), $r_F = 4 \times 10^{-15}\,\Omega.\text{m}^2$, $l^*_{sf,N} = 2\,\mu\text{m}$, and $d_N = 20\,\text{nm}$, and compare your findings with figure 67.

(g) Illustrate the effect of the spin conductivity mismatch on the spin current.

■ **Solutions**

(a) Spin information is to be transmitted between the F layers, across the N layer.

(b) $\Delta R/R^{(P)} = \beta^2/(1-\beta^2)$

(c) $\Delta R/R^{(P)} = 4\beta^2\left(\frac{r_F l^*_{sf,N}}{r_N d_N}\right)^2$

(d) As is, spin injection in a semiconductor is not efficient because $\frac{r_F l^*_{sf,N}}{r_N d_N} \ll 1$.

(e) - Adding interfacial spin resistances makes spin injection in a semiconductor possible.
- An insulating tunnel barrier gives a large spin-dependent resistance.

(f) Case (b), $\Delta R/R^{(P)} = 33\%$
Case (c), $\Delta R/R^{(P)} = 10^{-8} \sim 0$
Case (e), $\Delta R/R^{(P)} = 33\%$
Agrees with data obtained in figure 67.

(g) See figure 24.

6.5 Intra-Band Scattering – Intrinsic Damping (α_0) – chapters 2 and 4

■ **Introduction/Reminder**

The damping parameter α illustrates how fast magnetization relaxes toward equilibrium, when initially brought out-of-equilibrium (§4.4). The intrinsic intra-band mechanism accounts for relaxation *via* electron scattering.

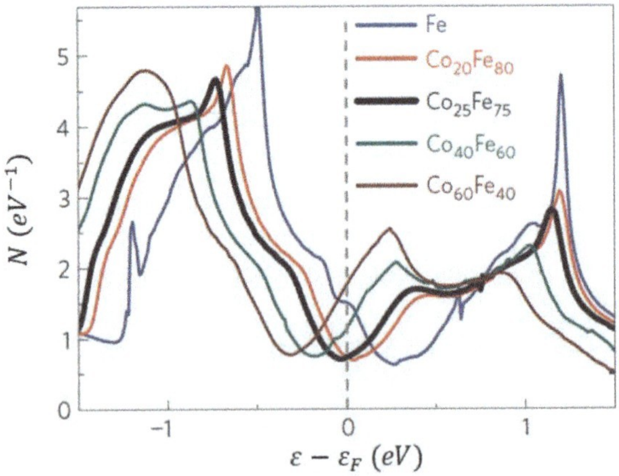

FIG. 68 – Calculated electron density of states (N) of Co_xFe_{100-x} alloys for several Co-concentrations. All alloy compositions are aligned to a common Fermi energy, ε_F, at zero energy. Adapted with permission from Springer Nature: M. A.W. Schoen *et al.*, *Nat. Phys.* **12**, 839 (2016). Copyright 2016.

■ Questions

(a) Give the typical equation describing magnetization dynamics, explain the different terms and illustrate how they contribute.
(b) Which other mechanism that you know of also depends on electron scattering? Please detail how.
(c) From the calculations shown in figure 68, plot the dependence of intrinsic damping (α_0) on the Co-concentration (x) in the Co_xFe_{100-x} ferromagnetic alloy. Please explain your plot.
(d) Which composition of the Co_xFe_{100-x} ferromagnetic alloy would you choose if you want to make a device with minimal fluctuations of magnetization after reversal, and why?

■ Solutions

(a) See equation (65) and corresponding text, §4.4.
(b) For example, the electronic mean free path is governed by the Fermi golden rule and therefore by the DOS at the Fermi level (see §2.4).
(c) The intrinsic intra-band mechanism accounts for relaxation *via* electron scattering, meaning that, α_0 is proportional to the DOS at $\varepsilon = \varepsilon_F$. From figure 68, we get:

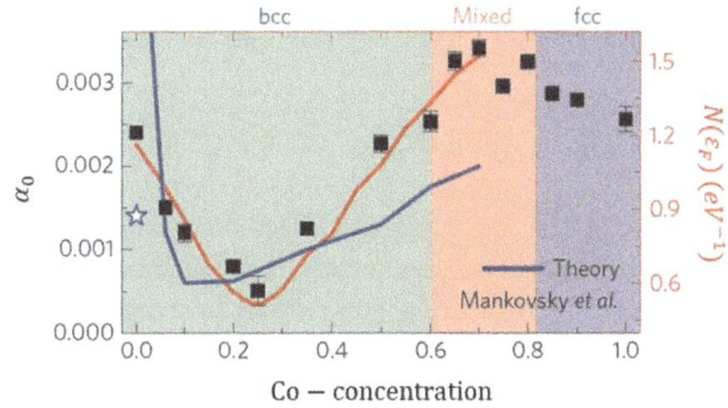

Adapted with permission from Springer Nature: M. A.W. Schoen *et al.*, *Nat. Phys.* **12**, 839 (2016). Copyright 2016.

(d) Large damping is suited for minimal fluctuations of magnetization, as it ensures fast relaxation. Based on (c), the largest intrinsic damping is obtained for the largest DOS at $\varepsilon = \varepsilon_F$, *i.e.* near the $Co_{60}Fe_{40}$ composition, from figure 68.

6.6 Spin Pumping (SP) and Inverse Spin Hall Effect (ISHE) – chapters 4 and 5

■ Introduction/Reminder

The spin pumping effect [Y. Tserkovnyak *et al.*, *Rev. Mod. Phys.* **77**, 1375 (2005)] refers to the ability of a magnetic material to generate a spin $\boldsymbol{s}$ current $\boldsymbol{J}_s^0$ when brought out-of-equilibrium (§4.4). The technique usually involves inducing resonance in a ferromagnetic (F) spin injector – *e.g.*, a NiFe layer – which is adjacent to a non-magnetic material (N) known as the spin sink – *e.g.* a Pt layer (figure 69). Spin pumping and spin transfer torque are reciprocal effect. An intuitive picture consists in comparing spin transfer torque to a water flow (spin current) moving the blades of a watermill (magnetization) and spin pumping to moving blades (magnetization) creating a water flow (spin current) (figure 41).

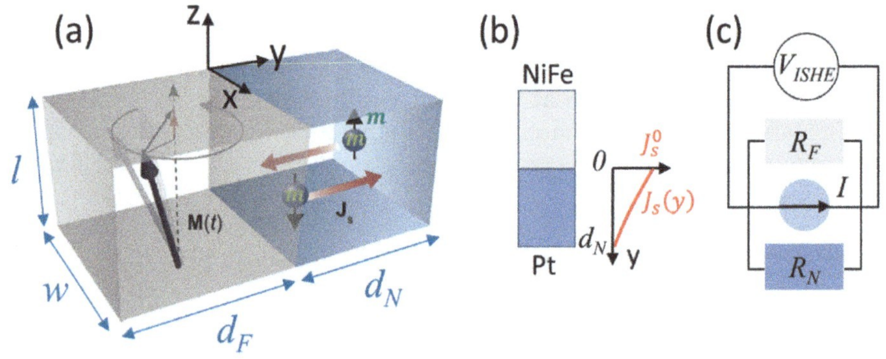

FIG. 69 – (a) Illustration of the spin pumping effect due to sustained out-of-equilibrium magnetization dynamics. (b) Illustration of spin current diffusion across the non-magnetic layer (here, Pt). (c) Equivalent circuit considering spin current to charge current conversion due to the inverse spin Hall effect in the non-magnetic layer. Also figure 42. Adapted with permission from AIP Publishing: K. Ando *et al.*, *J. Appl. Phys.* **109**, 103913 (2011). Copyright 2011.

■ Questions

(a) Because the system is out-of-equilibrium, spins accumulate at the F/N interface and diffuse across the N layer [figure 69b]. The time-varying spin density (nonequilibrium chemical-potential imbalance) can be written as follows: $\tilde{\mu}_s = \mu_s e^{i\omega t}$. In this part, you will calculate the y-dependence of the spin current in the N layer: $J_s(y)$.

(a1) Write the relation between J_s and μ_s.
This will be equation (a1).

(a2) Write the transport equation that regulates μ_s, and show that

$$\frac{d^2\mu_s}{dy^2} - \frac{1}{l_{sf}^{\#2}}\mu_s = 0, \tag{a2}$$

with $l_{sf}^{\#} = \frac{l_{sf}^*}{\sqrt{1+i\omega\tau_{sf}^*}}$ and $l_{sf}^* = \sqrt{D\tau_{sf}^*}$, where D is the diffusion constant, and τ_{sf}^* is the average spin-flip scattering rate in the N layer. Here, $l_{sf}^* = l_{sf,N}^*$ and $l_{sf}^{\#} = l_{sf,N}^{\#}$.

(a3) Is it realistic to consider that $l_{sf}^{\#} \sim l_{sf}^*$?

(a4) Write the boundary conditions at $y = 0$ and $y = d_N$.

(a5) Use the result of question (a4) and the fact that the solution of equation (a2) takes the following form: $\mu_s(y) = Ae^{y/l_{sf}^*} + Be^{-y/l_{sf}^*}$, to explicit $\mu_s(y)$ *vs.* y, d_N, l_{sf}^*, e, and ρ_N.

(a6) Use the result of question (a5) and equation (a1) to explicit $J_s(y)$ *vs.* y, d_N, l_{sf}^*, and J_s^0.
This will be equation (a6).

(b) Due to the inverse spin Hall effect (ISHE) (chapter 5), the spin current is converted in a transverse charge current J_e along x. The N layer then becomes a 'source' of charge current. The spin current to charge current conversion is expressed as $J_e(y) = \theta_{\text{SHE}} J_s(y)$, where θ_{SHE} is called the spin Hall angle.

(b1) Use equation (a6) and the indications above to explicit the average charge current density $J_e = \langle J_e(y)\rangle = \frac{1}{d_N}\int_0^{d_N} J_e(y)dy$ *vs.* y, d_N, l_{sf}^*, and J_s^0.

(b2) Plot the charge current I *vs.* d_N. Comment the trend for $d_N \ll l_{sf}^*$ and for $d_N \gg l_{sf}^*$.

(b3) The equivalent circuit of the F/N bilayer is illustrated in figure 69c. Explicit the electromotive force (voltage V_{ISHE}) due to the inverse spin Hall effect in the N layer induced by spin pumping in a F/N bilayer.

(b4) Calculate the value of the 'spin-charge conversion efficiency': $\theta_{\text{SHE}} l^*_{sf}$, considering that $d_N\sigma_N \gg d_F\sigma_F$, and for $V_{\text{ISHE}} = 4\ \mu\text{V}$, $d_N = 10$ nm, $l^*_{sf} =$ 3 nm, $\sigma_N = 4 \times 10^6\ \text{S.m}^{-1}$, $w = 0.5$ mm, and $J^0_s = 4.8 \times 10^5\ \text{A.m}^{-2}$.

(b5) A contribution to the spin Hall angle is due to extrinsic *sd* scattering on defects. Given this specific contribution – intrinsic contributions are not considered here – how would you proceed to increase the value of the spin Hall angle in a material? Will the options you propose improve the 'spin-charge conversion efficiency'?

(c) It is possible to show that the angular dependence of the inverse spin Hall voltage is

$$V_{\text{ISHE}} = \frac{w\theta_{\text{SHE}} l^*_{sf} \tanh\left(d_N/(2l^*_{sf})\right)}{d_N\sigma_N + d_F\sigma_F} \frac{e g_r^{\uparrow\downarrow}\gamma^2\left(\mu_0 h_{rf}\right)^2}{8\pi\alpha^2\omega} \sin(\theta_M)\overline{\Gamma}, \qquad \text{(c)}$$

with $\overline{\Gamma} = 2\omega \frac{\mu_0 M_S|\gamma|\sin^2(\theta_M) + \sqrt{(\mu_0 M_S\gamma\sin^2(\theta_M))^2 + 4\omega^2}}{(\mu_0 M_S\gamma\sin^2(\theta_M))^2 + 4\omega^2}$, w the width of the layers, θ_{SHE} the spin-Hall angle, h_{rf} the *rf* excitation field used to reach F resonance, and ω the resonance angular frequency. θ_M is defined in figure 70. $\overline{\Gamma}$ is a parameter accounting for the trajectory of the magnetization precession in the (xy) plane. It can be viewed as the elliptic- to circular-trajectory ratio. $g_r^{\uparrow\downarrow}$ is the real part of the spin mixing conductance per unit area per quantum conductance per spin channel, accounting for the ability of the F/N interface and N layer to absorb the spin component along $\boldsymbol{M} \times (\boldsymbol{m} \times \boldsymbol{M})$.

(c1) Show that $\overline{\Gamma}$ is a dimensionless parameter.

(c2) What is the trajectory of the magnetization precession for $\theta_M = 0$? Why is it so? And should it always be the case?

(c3) The angular(θ_H)-dependence of the magnetization's tilt θ_M deduced from experimental data for the case of a 8 nm-thick NiFe film is given in figure 70. θ_H is the angle between the applied magnetic field $\boldsymbol{H}$ and the normal to the sample's surface $\boldsymbol{y}$. Comment this behaviour: what governs it? Use equation (c) and figure 70, to hand-sketch the θ_H-dependence of V_{ISHE}. Comment the symmetry of V_{ISHE} with $\boldsymbol{H}$.

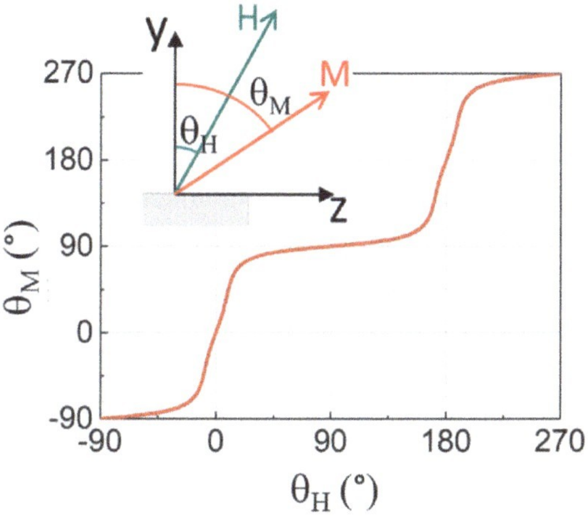

FIG. 70 – Typical angular(θ_H)-dependence of the magnetization's tilt θ_M. From O. Gladii *et al.*, *Phys. Rev. B* **100**, 174409 (2019).

■ Solutions

(a1) The relation between J_s and μ_s is

$$J_s = \frac{1}{2e\rho_N}\frac{d\mu_s}{dy} \tag{a1}$$

Note: take $\sigma^\uparrow = \sigma^\downarrow = \sigma_N/2$

(a2) The time-variation of the spin density $\tilde{\mu}_s$ results in spin diffusion, which is balanced by spin scattering. The transport equation becomes

$$\frac{d\tilde{\mu}_s}{dt} = D\frac{d^2\tilde{\mu}_s}{dy^2} - \frac{\tilde{\mu}_s}{\tau^*_{sf}},$$

$$i\omega\mu_s = D\frac{d^2\mu_s}{dy^2} - \frac{\mu_s}{\tau^*_{sf}},$$

$$\frac{d^2\mu_s}{dy^2} - \frac{1}{l^{\#\,2}_{sf}}\mu_s = 0 \tag{a2}$$

with $l^{\#}_{sf} = \frac{l^*_{sf}}{\sqrt{1+i\omega\tau^*_{sf}}}$ and $l^*_{sf} = \sqrt{D\tau^*_{sf}}$, where D is the diffusion constant and τ^*_{sf} is the average spin-flip scattering rate.

(a3) We have $l_{sf}^{\#} = \frac{l_{sf}^{*}}{\sqrt{1+i\omega\tau_{sf}^{*}}}$ with $\omega\tau_{sf}^{*} = 2\pi f\tau_{sf}^{*}$. The typical value of the resonance frequency for a ferromagnet is $f \sim 10$ GHz (with $\mu_0 H \sim 0.1$ T), and the typical value of spin-flip scattering time is $\tau_{sf}^{*} \sim 10$ fs, so $\omega\tau_{sf}^{*} \sim 10^{11}\times 10^{-14} \ll 1$ => $l_{sf}^{\#} \sim l_{sf}^{*}$. Note that this condition is no longer true in the THz regime, for example for antiferromagnetic resonance. In the following, we consider that $l_{sf}^{\#} = l_{sf}^{*}$.

(a4) At the boundaries, we have:
for $y = 0$, $J_s = \frac{1}{2e\rho_N}\frac{d\mu_s}{dy}(0) = J_s^0$,
for $y = d_N$, $J_s = \frac{1}{2e\rho_N}\frac{d\mu_s}{dy}(d_N) = 0$.

(a5) The solution of equation (a2) takes the form

$$\mu_s(y) = Ae^{y/l_{sf}^{*}} + Be^{-y/l_{sf}^{*}},$$

for $y = 0$, $\frac{1}{2e\rho_N}\frac{d\mu_s}{dy} = J_s^0$ => $A - B = 2e\rho_N J_s^0 l_{sf}^{*}$
for $y = d_N$, $\frac{1}{2e\rho_N}\frac{d\mu_s}{dy} = 0$ => $B = Ae^{2d_N/l_{sf}^{*}}$

$$\mu_s(y) = 2e\rho_N J_s^0 \frac{l_{sf}^{*}}{1 - e^{2d_N/l_{sf}^{*}}}\left(e^{y/l_{sf}^{*}} + e^{-y/l_{sf}^{*}}e^{2d_N/l_{sf}^{*}}\right)$$

$$\mu_s(y) = 2e\rho_N J_s^0 \frac{e^{-d_N/l_{sf}^{*}}}{e^{-d_N/l_{sf}^{*}}}\frac{l_{sf}^{*}}{1 - e^{2d_N/l_{sf}^{*}}}\left(e^{y/l_{sf}^{*}} + e^{-y/l_{sf}^{*}}e^{2d_N/l_{sf}^{*}}\right)$$

$$\mu_s(y) = -2e\rho_N J_s^0 l_{sf}^{*}\frac{\cosh\left((y - d_N)/l_{sf}^{*}\right)}{\sinh\left(d_N/l_{sf}^{*}\right)}$$

(a6) Using the result of question (a5) and equation (a1), we obtain

$$J_s(y) = -J_s^0 \frac{\sinh\left((y - d_N)/l_{sf}^{*}\right)}{\sinh\left(d_N/l_{sf}^{*}\right)} \qquad \text{(a6)}$$

(b1)
$$J_e = \langle J_e(y)\rangle = \frac{1}{d_N}\int_0^{d_N} \theta_{\mathrm{SHE}} J_s(y)\,dy = -\frac{1}{d_N}\int_0^{d_N} \theta_{\mathrm{SHE}} J_s^0 \frac{\sinh\left((y-d_N)/l_{sf}^*\right)}{\sinh\left(d_N/l_{sf}^*\right)}\,dy$$

$$J_e = -\theta_{\mathrm{SHE}} J_s^0 \frac{1}{d_N}\frac{1}{\sinh\left(d_N/l_{sf}^*\right)}\int_0^{d_N} \sinh\left((y-d_N)/l_{sf}^*\right)dy$$

$$J_e = -\theta_{\mathrm{SHE}} J_s^0 \frac{l_{sf}^*}{d_N}\frac{1-\cosh\left(d_N/l_{sf}^*\right)}{\sinh\left(d_N/l_{sf}^*\right)}$$

$$= -\theta_{\mathrm{SHE}} J_s^0 \frac{l_{sf}^*}{d_N}\frac{1-2\sinh^2\left(d_N/\left(2l_{sf}^*\right)\right)+1}{2\sinh\left(d_N/\left(2l_{sf}^*\right)\right)\cosh\left(d_N/\left(2l_{sf}^*\right)\right)}$$

$$J_e = \theta_{\mathrm{SHE}} \frac{l_{sf}^*}{d_N}\tanh\left(\frac{d_N}{2l_{sf}^*}\right) J_s^0$$

$$J_e = \theta_{\mathrm{SHE}} \frac{l_{sf}^*}{d_N}\tanh\left(\frac{d_N}{2l_{sf}^*}\right) J_s^0$$

(b2) $I = J_e l d_N = \theta_{\mathrm{SHE}} l l_{sf}^* \tanh\left(\frac{d_N}{2l_{sf}^*}\right) J_s^0$

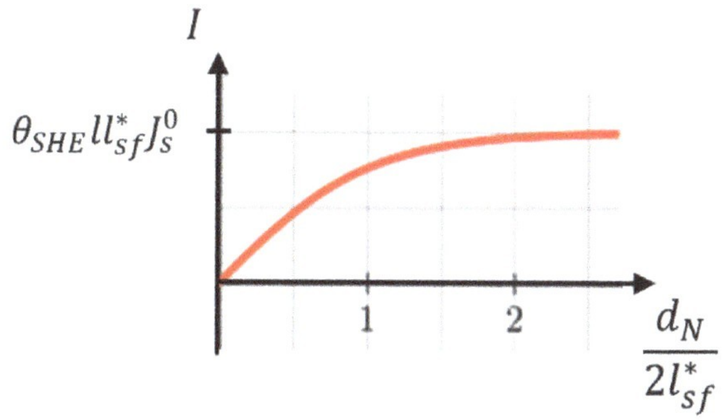

For $d_N \ll l_{sf}^*$, $I \sim \theta_{\mathrm{SHE}} l \frac{d_N}{2} J_s^0$. Spins are still polarized and get converted efficiently in the N layer. The thicker the layer, the more spins are converted.
For $d_N \gg l_{sf}^*$, $I = \theta_{\mathrm{SHE}} l l_{sf}^* J_s^0$. The signal levels out. The part of the N layer in contact with the F layer ($d_N < 2l_{sf}^*$) converts spins. Above $d_N \sim 2l_{sf}^*$ spin-charge conversion becomes inefficient because the spins are depolarized.

(b2) $$V_{\text{ISHE}} = \frac{R_F R_N}{R_{F+} R_N} I = \frac{\frac{\rho_F w}{l d_F} \frac{\rho_N w}{l d_N}}{\frac{\rho_F w}{l d_F} + \frac{\rho_N w}{l d_N}} J_e l d_N$$

$$V_{\text{ISHE}} = \frac{w d_N}{d_N \sigma_N + d_F \sigma_F} J_e$$

$$V_{\text{ISHE}} = \theta_{\text{SHE}} \frac{w l^*_{sf}}{d_N \sigma_N + d_F \sigma_F} \tanh\left(\frac{d_N}{2 l^*_{sf}}\right) J_s^0$$

(b3) The Hall angle is $\theta_{\text{SHE}} \sim 6\%$ and the 'spin-charge conversion efficiency' is $\theta_{\text{SHE}} l^*_{sf} \sim 0.18$ nm [J. C. Rojas-Sanchez *et al.*, *Phys. Rev. Lett.* **112**, 106602 (2014)].

(b4) Increasing the number of scattering centers will increase the extrinsic contribution to θ_{SHE}, because for this extrinsic contribution $\theta_{\text{SHE}} \propto \frac{1}{l^*_{sf}}$, and increasing the number of scatterers will reduce l^*_{sf}. This solution is however not appropriate to increase the 'spin-charge conversion efficiency' because $\theta_{\text{SHE}} l^*_{sf}$ will remain unchanged.
Increasing the scattering efficiency by using heavier materials, with a large atomic number Z, is another solution to increase θ_{SHE}. Because spin-orbit interactions roughly vary as Z^4, this solution will result in a large increase in 'spin-charge conversion efficiency'. Note however that other effects contribute to θ_{SHE} and the picture presented above is certainly more complicated. In particular, intrinsic contributions do matter, see chapter 5.

(c1) $$\overline{\Gamma} = 2\omega \frac{\mu_0 M_S |\gamma| \sin^2(\theta_M) + \sqrt{(\mu_0 M_S \gamma \sin^2(\theta_M))^2 + 4\omega^2}^{\,2}}{(\mu_0 M_S \gamma \sin^2(\theta_M))^2 + 4\omega^2}$$
γ is in unit of Hz.T^{-1}; $\mu_0 M_S$ is in T; and ω is in Hz, so we can conclude that $\overline{\Gamma}$ is a dimensionless parameter.

(c2) For $\theta_M = 0$, $\overline{\Gamma} = 1$. By definition, the magnetization trajectory in the (x,y) plane is thus circular. In this case, the magnetization rotates about the out-of-plane direction. If there is no in-plane anisotropy, it is expected that the trajectory is circular. It would not be the case in the presence of an in-plane anisotropy.

(c3) The magnetization is mostly in-plane, whatever the angle of the applied field. This is because of the demagnetizing field.

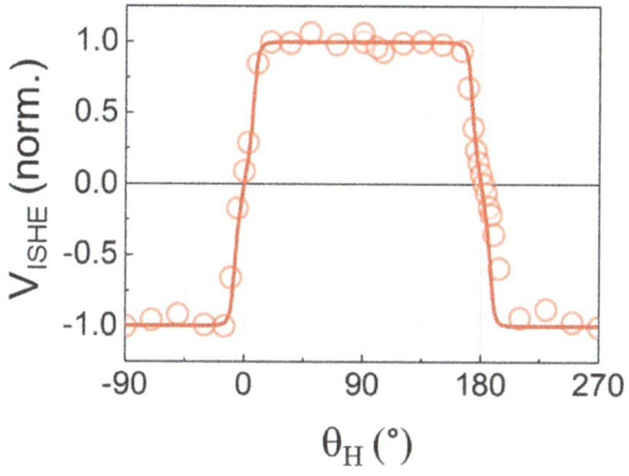

From O. Gladii *et al.*, *Phys. Rev. B* **100**, 174409 (2019).

V_{ISHE} is an odd function of $\boldsymbol{H}$. Angular-dependent symmetries are useful to disentangle the inverse spin Hall effect and several others that may occur concurrently [M. Harder *et al.*, *Phys. Rep.* **661**, 1 (2016)]. When two effects share the same angular-dependent symmetries, frequency-, temperature-, stacking order- etc. dependences are used to unravel the contributions.

6.7 Spin Pumping (SP) – Additional Damping (α_p) – chapter 4

■ Introduction/Reminder

This exercise is related to the previous one. Its purpose is to demonstrate that spin pumping (§4.4) generates a damping term and to calculate this term. Magnetization dynamics (figure 69a) is described by the LLG equation. In the frame of spin pumping, one takes into account a term accounting for a loss of spin current due to the transfer of angular momentum at the F/N interface, at $y = 0$. The LLG equation takes the form [R. Igushi *et al.*, *J. Phys. Soc. Jap.* **86**, 011003 (2017)].

$$\frac{d\boldsymbol{M}}{dt} = -|\gamma|\boldsymbol{M} \times \left(\mu_0\boldsymbol{H}_{\text{eff}} + \mu_0\boldsymbol{h}_{rf}\right) + \frac{\alpha_0}{M_S}\boldsymbol{M} \times \frac{d\boldsymbol{M}}{dt} - |\gamma|\boldsymbol{T}_{\text{STT}}^{\text{SP}}$$

The torque $\boldsymbol{T}_{\text{STT}}^{\text{SP}}$ can be obtained based on the conservation law of total angular momentum, which local formulation is $\dfrac{d\boldsymbol{M}}{dt} = |\gamma|\boldsymbol{\nabla} \cdot \left(\dfrac{\hbar}{2e}\boldsymbol{J}_s\right) = -|\gamma|\boldsymbol{T}$ [equation (75) and corresponding text]. From figure 69a, we see that, when incoming electrons provide $\boldsymbol{m}$ to the magnetization $\boldsymbol{M}$, this latter orients towards $\boldsymbol{m}$, and $\frac{d\boldsymbol{M}}{dt} > 0$, in agreement with the ideas discussed in chapter 4.

The LLG equation then becomes

$$\frac{d\boldsymbol{M}}{dt} = \gamma \boldsymbol{M} \times \left(\mu_0 \boldsymbol{H}_{\text{eff}} + \mu_0 \boldsymbol{h}_{rf}\right) + \frac{\alpha_0}{M_S} \boldsymbol{M} \times \frac{d\boldsymbol{M}}{dt} + \frac{|\gamma|}{d_F} \frac{\hbar}{2e} \boldsymbol{J}_s^0$$

with $\boldsymbol{J}_s^0$ the spin current at the interface [Y. Tserkovnyak *et al.*, *Phys. Rev. B* **66**, 224403 (2002)]

$$\boldsymbol{J}_s^0 = \boldsymbol{J}_s^{\text{pump}} - \boldsymbol{J}_s^{\text{back}} = \frac{e}{2\pi M_S^2} g_{r,\text{eff}}^{\uparrow\downarrow} \boldsymbol{M} \times \frac{d\boldsymbol{M}}{dt} + \frac{e}{2\pi M_S} g_{i,\text{eff}}^{\uparrow\downarrow} \frac{d\boldsymbol{M}}{dt}$$

We recall that the definitions and uses of $\boldsymbol{J}_s$ and γ may vary, resulting in a factor $\frac{\hbar}{2e}$ difference and ± 1 or $\pm\mu_0$, respectively depending on the articles and books consulted. In this textbook, the spin current density is $\boldsymbol{J}_s = -\left(\boldsymbol{j}^{\uparrow} - \boldsymbol{j}^{\downarrow}\right)$, expressed in A.m^{-2} and the gyromagnetic ratio $\gamma < 0$ is taken as $-|\gamma|$.

■ Questions

(a) Inject $\boldsymbol{J}_s^0$ in the LLG equation and show that the total damping takes the form $\alpha = \alpha_0 + \alpha_p$. Give the expression of α_p. Note: we will make the approximation that the imaginary part of the effective spin mixing conductance $g_{i,\text{eff}}^{\uparrow\downarrow}$ averages out (FL term page 58).

(b) For this question, we consider that:
- the initial spin current pumped writes as follows:

$$\boldsymbol{J}_s^{\text{pump}} = \frac{e}{2\pi M_S^2} g_r^{\uparrow\downarrow} \boldsymbol{M} \times \frac{d\boldsymbol{M}}{dt} + \frac{e}{2\pi M_S} g_i^{\uparrow\downarrow} \frac{d\boldsymbol{M}}{dt}$$

- the backflow spin current at the interface, at $y = 0$, relates to the transfer of angular momentum from spin accumulation in the non-magnetic layer to the ferromagnet magnetization as

$$\boldsymbol{J}_s^{\text{back}} = -\frac{e}{2\pi M_S^2} g_r^{\uparrow\downarrow} \boldsymbol{M} \times \left(\boldsymbol{M} \times \frac{\boldsymbol{\mu}_s^0}{\hbar}\right) - \frac{e}{2\pi M_S} g_i^{\uparrow\downarrow} \boldsymbol{M} \times \frac{\boldsymbol{\mu}_s^0}{\hbar}$$

Note: we will also make the approximation that the imaginary part of the spin mixing conductance $g_i^{\uparrow\downarrow}$ averages out.

(b1) Express $g_{r,\text{eff}}^{\uparrow\downarrow}$ as a function of $g_r^{\uparrow\downarrow}$ when $\boldsymbol{J}_s^{\text{back}} = 0$.

(b2) Based on the drift-diffusion equation for spin accumulation, give the expression of $\boldsymbol{\mu}_s^0$, then deduce the expression of $\boldsymbol{J}_s^{\text{back}}$ and finally demonstrate that $g_{r,\text{eff}}^{\uparrow\downarrow} = \left[\frac{1}{g_r^{\uparrow\downarrow}} + \frac{2l_{sf}^*}{G_0\sigma_N}\tanh^{-1}\left(\frac{d_N}{l_{sf}^*}\right)\right]^{-1}$, where $l_{sf}^* = l_{sf,N}^*$.

(b3) Comment this last formula when $l_{sf}^* \gg d_N$.

■ Solutions

(a) Based on the introduction/reminder part, and neglecting $g_{i,\text{eff}}^{\uparrow\downarrow}$, the LLG equations becomes

$$\frac{d\boldsymbol{M}}{dt} = -|\gamma|\boldsymbol{M} \times (\mu_0\boldsymbol{H}_{\text{eff}} + \mu_0\boldsymbol{h}_{rf}) + \frac{1}{M_S}\left(\alpha_0 + \frac{|\gamma|}{M_S d_F}\frac{\hbar}{4\pi}g_{r,\text{eff}}^{\uparrow\downarrow}\right)\boldsymbol{M} \times \frac{d\boldsymbol{M}}{dt}$$

Due to spin pumping (sinking), the damping of the ferromagnetic layer therefore writes

$$\alpha = \alpha_0 + \alpha_p$$

with

$$\alpha_p = \frac{|\gamma|}{M_S d_F}\frac{\hbar}{4\pi}g_{r,\text{eff}}^{\uparrow\downarrow}$$

(b1) If one neglects $\boldsymbol{J}_s^{\text{back}}$, then $\boldsymbol{J}_s^0 = \boldsymbol{J}_s^{\text{pump}}$ and

$$g_{r,\text{eff}}^{\uparrow\downarrow} = g_r^{\uparrow\downarrow}$$

(b2) The expression of $\boldsymbol{\mu}_s^0$ can be obtained by solving the drift-diffusion equation for spin accumulation in the non-magnetic layer, for $y = 0$

$$\frac{d^2\mu_s}{dy^2} - \frac{1}{l_{sf}^{*2}}\mu_s = 0$$

At the boundaries, we have the following conditions:
for $y = 0$, $\frac{1}{2e\rho_N}\frac{d\mu_s}{dy}(0) = J_s^0$, and for $y = d_N$, $\frac{1}{2e\rho_N}\frac{d\mu_s}{dy}(d_N) = 0$.
The solution of the drift-diffusion equation takes the form.

$$\mu_s(y) = Ae^{y/l_{sf}^*} + Be^{-y/l_{sf}^*},$$

for $y=0$, $\frac{1}{2e\rho_N}\frac{d\mu_s}{dy}(0)=J_s^0 \Rightarrow A-B=2e\rho_N J_s^0 l_{sf}^*$
for $y=d_N$, $\frac{1}{2e\rho_N}\frac{d\mu_s}{dy}(d_N)=0 \Rightarrow B=Ae^{2d_N/l_{sf}^*}$

$$\mu_s(y)=2e\rho_N J_s^0\frac{l_{sf}^*}{1-e^{2d_N/l_{sf}^*}}\left(e^{y/l_{sf}^*}+e^{-y/l_{sf}^*}e^{2d_N/l_{sf}^*}\right)$$

$$\mu_s(y)=2e\rho_N J_s^0\frac{e^{-d_N/l_{sf}^*}}{e^{-d_N/l_{sf}^*}}\frac{l_{sf}}{1-e^{2d_N/l_{sf}^*}}\left(e^{y/l_{sf}^*}+e^{-y/l_{sf}^*}e^{2d_N/l_{sf}^*}\right)$$

$$\mu_s(y)=-2e\rho_N J_s^0 l_{sf}^*\frac{\cosh\left((y-d_N)/l_{sf}^*\right)}{\sinh\left(d_N/l_{sf}^*\right)}$$

$$\mu_s^0=-2e\rho_N J_s^0 l_{sf}^*\coth\left(d_N/l_{sf}^*\right)$$

$$\boldsymbol{\mu}_s^0=-\frac{2el_{sf}^*}{\sigma_N}\coth\left(d_N/l_{sf}^*\right)\boldsymbol{J}_s^0$$

Note: see also answers to question (a) – a1 to a5 – in the previous exercise. Neglecting the imaginary part, we have

$$\boldsymbol{J}_s^{\text{back}}=\frac{e}{2\pi M_S^2}g_r^{\uparrow\downarrow}\boldsymbol{M}\times\left(\frac{\boldsymbol{\mu}_s^0}{\hbar}\times\boldsymbol{M}\right)$$

$$\boldsymbol{J}_s^{\text{back}}=\frac{e}{2\pi M_S^2}g_r^{\uparrow\downarrow}\boldsymbol{M}\times\left(\boldsymbol{M}\times\frac{2el_{sf}^*}{\hbar\sigma_N}\coth\left(d_N/l_{sf}^*\right)\boldsymbol{J}_s^0\right)$$

From $\boldsymbol{J}_s^0=\boldsymbol{J}_s^{\text{pump}}-\boldsymbol{J}_s^{\text{back}}$, based on the expression of $\boldsymbol{J}_s^0$, $\boldsymbol{J}_s^{\text{pump}}$ and $\boldsymbol{J}_s^{\text{back}}$, and neglecting the imaginary parts, one obtains

$$\frac{e}{2\pi M_S^2}g_{r,\text{eff}}^{\uparrow\downarrow}\boldsymbol{M}\times\frac{d\boldsymbol{M}}{dt}=\frac{e}{2\pi M_S^2}g_r^{\uparrow\downarrow}\boldsymbol{M}\times\frac{d\boldsymbol{M}}{dt}+\frac{e}{2\pi M_S^2}g_r^{\uparrow\downarrow}\boldsymbol{M}\times\left(\boldsymbol{M}\times\frac{2el_{sf}^*}{\hbar\sigma_N}\coth\left(d_N/l_{sf}^*\right)\boldsymbol{J}_s^0\right)$$

$$\frac{e}{2\pi M_S^2}g_{r,\text{eff}}^{\uparrow\downarrow}\boldsymbol{M}\times\frac{d\boldsymbol{M}}{dt}=\frac{e}{2\pi M_S^2}g_r^{\uparrow\downarrow}\boldsymbol{M}\times\frac{d\boldsymbol{M}}{dt}+\frac{e}{2\pi M_S^2}g_r^{\uparrow\downarrow}\boldsymbol{M}\times\left(\boldsymbol{M}\times\frac{2el_{sf}^*}{\hbar\sigma_N}\coth\left(d_N/l_{sf}^*\right)\left(\frac{e}{2\pi M_S^2}g_{r,\text{eff}}^{\uparrow\downarrow}\boldsymbol{M}\times\frac{d\boldsymbol{M}}{dt}\right)\right)$$

$$g_{r,\text{eff}}^{\uparrow\downarrow}\boldsymbol{M}\times\frac{d\boldsymbol{M}}{dt}=g_r^{\uparrow\downarrow}\boldsymbol{M}\times\frac{d\boldsymbol{M}}{dt}+g_r^{\uparrow\downarrow}\boldsymbol{M}$$
$$\times\left(\boldsymbol{M}\times\frac{2el_{sf}^*}{\hbar\sigma_N}\coth\left(\frac{d_N}{l_{sf}^*}\right)\left(\frac{e}{2\pi M_S^2}g_{r,\text{eff}}^{\uparrow\downarrow}\boldsymbol{M}\times\frac{d\boldsymbol{M}}{dt}\right)\right)$$
$$g_{r,\text{eff}}^{\uparrow\downarrow}=g_r^{\uparrow\downarrow}-g_r^{\uparrow\downarrow}\frac{2el_{sf}^*}{\hbar\sigma_N}\coth\left(\frac{d_N}{l_{sf}^*}\right)\frac{e}{2\pi}g_{r,\text{eff}}^{\uparrow\downarrow}$$
$$g_{r,\text{eff}}^{\uparrow\downarrow}=g_r^{\uparrow\downarrow}-g_r^{\uparrow\downarrow}\frac{2e^2l_{sf}^*}{h\sigma_N}\coth\left(d_N/l_{sf}^*\right)g_{r,\text{eff}}^{\uparrow\downarrow}$$
$$g_{r,\text{eff}}^{\uparrow\downarrow}\tanh\left(d_N/l_{sf}^*\right)=g_r^{\uparrow\downarrow}\tanh\left(d_N/l_{sf}^*\right)-g_r^{\uparrow\downarrow}\frac{2e^2l_{sf}^*}{h\sigma_N}g_{r,\text{eff}}^{\uparrow\downarrow}$$
$$g_{r,\text{eff}}^{\uparrow\downarrow}\left[\tanh\left(\tfrac{d_N}{l_{sf}^*}\right)+g_r^{\uparrow\downarrow}\frac{2e^2l_{sf}^*}{h\sigma_N}\right]=g_r^{\uparrow\downarrow}\tanh\left(d_N/l_{sf}^*\right)$$
$$g_{r,\text{eff}}^{\uparrow\downarrow}=\frac{g_r^{\uparrow\downarrow}\tanh\left(d_N/l_{sf}^*\right)}{\tanh\left(\frac{d_N}{l_{sf}^*}\right)+g_r^{\uparrow\downarrow}\frac{2e^2l_{sf}^*}{h\sigma_N}}$$
$$g_{r,\text{eff}}^{\uparrow\downarrow}=\left[\frac{1}{g_r^{\uparrow\downarrow}}+\frac{2l_{sf}^*}{G_0\sigma_N}\tanh^{-1}\left(\frac{d_N}{l_{sf}^*}\right)\right]^{-1}$$

By comparing the results of questions (b1) and (b2), we conclude that taking into account the backflow spin current renormalizes the spin mixing conductance.

We also see an expression that looks like resistors in series. More information on the spin-circuit representation of spin pumping can be found in K. Roy *et al.*, *Phys. Rev. Appl.* **8**, 011001 (2017).

(b3) When $l_{sf}^*\gg d_N$, $g_{r,\text{eff}}^{\uparrow\downarrow}=0$. Indeed, no spin has been absorbed and all spins have been reflected: $\boldsymbol{J}_s^{\text{pump}}=\boldsymbol{J}_s^{\text{back}}$, and thus $\boldsymbol{J}_s^0=0$. In other words, no net transfer of angular momentum took place.

6.8 Spin Hall Magnetoresistance (SMR) – chapters 2–4

■ **Introduction/Reminder**

This exercise is based on a spintronic effect called the spin Hall magnetoresistance (SMR), published in H. Nakayama *et al.*, *Phys. Rev. Lett.* **110**, 206601 (2013). It illustrates knowledge gained in chapters 2–4. A bilayer is considered (figure 71). It is made of a metallic heavy 'non-magnet' (HM), like Pt, and an insulating magnetic layer, like YIG (F) (YIG = Yttrium iron garnet, $Y_3Fe_5O_{12}$).

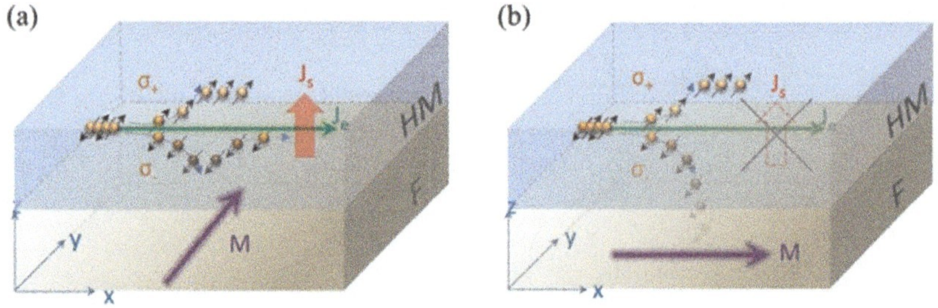

FIG. 71 – Illustration of the spin Hall magnetoresistance effect. From S.-Y. Huang *et al.*, *Sci. Rep.* **8**, 108 (2018) – CC BY Licence.

A charge current $\boldsymbol{J}_e$ is passed along the $\widehat{\boldsymbol{x}}$ direction in the HM. The spin Hall effect (SHE) converts this charge current in a transverse spin current $\boldsymbol{J}_s$. The amplitude of the spin current is controlled by the orientation of the F magnetization $\boldsymbol{M} = M_S\widehat{\boldsymbol{M}}$, due to transfer of angular momentum *via* spin transfer torque. We note α, the angle between $\widehat{\boldsymbol{M}}$ and $\widehat{\boldsymbol{y}}$.

In the illustrations of figure 71, $\boldsymbol{J}_s = \boldsymbol{J}_s^y = J_{sz}^y(z)\widehat{\boldsymbol{z}}$, meaning that the spin current is made of spins polarized along $\widehat{\boldsymbol{y}}$ only (first part of the equation) and that those spins flow along $\widehat{\boldsymbol{z}}$ only (second part), in a way that is z-dependent.

For $\alpha = \pi/2$, figure 71a, when $\widehat{\boldsymbol{M}} = \widehat{\boldsymbol{y}}$, the polarization of the spin current in the HM and the magnetization of the F are parallel. As a result, there is no transfer of angular momentum. Spin accumulation in the HM is $\boldsymbol{\mu}_s = \mu_{sy}(z)\widehat{\boldsymbol{y}}$, meaning that the spin accumulation vector has a component along $\widehat{\boldsymbol{y}}$ only. It creates a backflow spin current, which compensates for the triggering spin current. The resistivity of the HM is low.

For $\alpha = 0$, figure 71b, when $\widehat{\boldsymbol{M}} = \widehat{\boldsymbol{x}}$, the polarization of the spin current in the HM and the F magnetization are perpendicular. As a result, there is transfer of angular momentum and the spin current is partially absorbed by F. Spin accumulation in the HM still writes $\boldsymbol{\mu}_s = \mu_{sy}(z)\widehat{\boldsymbol{y}}$. It is smaller compared to the case $\alpha = \pi/2$. The resistivity of the HM is high.

The purpose of this exercise is to calculate the change in resistivity with α, *i.e.* with the orientation of $\widehat{\boldsymbol{M}}$, which can be changed, in practice, by use of an external magnetic field $\boldsymbol{H}$.

The general Ohm's law in HM is

$$\begin{pmatrix} \boldsymbol{J}_e \\ \boldsymbol{J}_s^x \\ \boldsymbol{J}_s^y \\ \boldsymbol{J}_s^z \end{pmatrix} = \sigma_N \begin{pmatrix} 1 & \theta_{\mathrm{SHE}}\widehat{\boldsymbol{x}}\times & \theta_{\mathrm{SHE}}\widehat{\boldsymbol{y}}\times & \theta_{\mathrm{SHE}}\widehat{\boldsymbol{z}}\times \\ \theta_{\mathrm{SHE}}\widehat{\boldsymbol{x}}\times & 1 & 0 & 0 \\ \theta_{\mathrm{SHE}}\widehat{\boldsymbol{y}}\times & 0 & 1 & 0 \\ \theta_{\mathrm{SHE}}\widehat{\boldsymbol{z}}\times & 0 & 0 & 1 \end{pmatrix} \begin{pmatrix} \boldsymbol{E} \\ \boldsymbol{\nabla}\mu_{sx}/(2e) \\ \boldsymbol{\nabla}\mu_{sy}/(2e) \\ \boldsymbol{\nabla}\mu_{sz}/(2e) \end{pmatrix}$$

Here, $\boldsymbol{E} = E_x\widehat{\boldsymbol{x}}$ is the applied electric field, $-e$ is the electron charge, σ_N is the HM conductivity, and θ_{SHE} is the spin Hall angle of the HM layer.

■ Questions

(a) We first start by treating the case $\alpha = \pi/2$ [until question (f)]. Give the expression of $\begin{pmatrix} \boldsymbol{J}_e \\ \boldsymbol{J}_s^y \end{pmatrix}$ as a function of $\begin{pmatrix} \widehat{\boldsymbol{x}} \\ \widehat{\boldsymbol{z}} \end{pmatrix}$.

(b) Use the diffusion equation for spin accumulation to find the expression of $\mu_{sy}(z)$. For this part, consider (i) a semi-infinite layer and (ii) the following boundary condition at $z = 0$, *i.e.*, at the interface between HM and F

$$\boldsymbol{J}_s^y(0) = -\left[\frac{e}{2\pi} g_r^{\uparrow\downarrow} \widehat{\boldsymbol{M}} \times \left(\widehat{\boldsymbol{M}} \times \frac{\boldsymbol{\mu}_s^0}{\hbar}\right)\right]\widehat{\boldsymbol{y}}$$

(c) What is the physical meaning of the above mentioned boundary condition?

(d) Using questions (a) and (b), give the expression of $\boldsymbol{J}_e = J_e(z)\widehat{\boldsymbol{x}}$

(e) Write $J_e(z)$ in the form $J_e(z) = J_e^0 + J_e^0 f(z)$ and use the approximation $J_e = J_e^0 + J_e^0 \frac{1}{d_N}\int_0^\infty f(z)dz$, to find the 'average' charge current. d_N is the thickness of HM.

(f) Set $J_e = \sigma_{xx}E_x$, and expand to leading order in θ_{SHE}^2 to find the expression of $\rho_{xx} = \rho_{xx,\alpha=\pi/2}$.

(g) For $\alpha = 0$, go over questions (a)–(f) again and give the expression of $\rho_{xx} = \rho_{xx,\alpha=0}$.

(h) The expression of the spin Hall magnetoresistance (SMR) is defined as $\Delta\rho_{xx} = \rho_{xx,\alpha=0} - \rho_{xx,\alpha=\pi/2} = \rho_0 - \rho_{\pi/2}$. Verify that $\Delta\rho_{xx} = \rho_N \theta_{\text{SHE}}^2 \frac{l_{sf}^*}{d_N} \frac{l_{sf}^* g_r^{\uparrow\downarrow} G_0}{\sigma_N + l_{sf}^* g_r^{\uparrow\downarrow} G_0}$.

(i) The actual full angular dependence of the SMR signal is given as follows (see figure 72 for the definition of the angles):

$$\rho_{xx,\text{SMR}}(\alpha) = \rho_0 + \Delta\rho_{xx}\cos^2\alpha$$

$$\rho_{xx,\text{SMR}}(\beta) = \rho_0 + \Delta\rho_{xx}\sin^2\beta$$

$$\rho_{xx,\text{SMR}}(\gamma) = \rho_0$$

Explain why $\rho_{xx,\text{SMR}}(\gamma) = \rho_0$ is independent of γ.

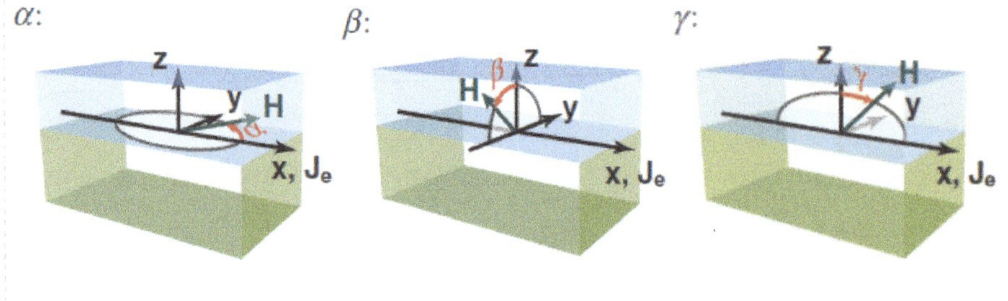

FIG. 72 – Illustration of the angles considered. Adapted with permission from the American Physical Society: H. Nakayama *et al.*, *Phys. Rev. Lett.* **110**, 206601 (2013). Copyright 2013.

(j) We now consider that the HM layer is slightly magnetized due to proximity effects, meaning that it now consists of a metallic ferromagnet with magnetization $\boldsymbol{M_{HM}} = M_{HM}\widehat{\boldsymbol{M}}_{\boldsymbol{HM}}$. This magnetization does not influence the SMR signal. However, it creates an additional signal due to anisotropic magnetoresistance (AMR).

(j1) Explain what AMR is.

(j2) Write how AMR depends on the orientation of the external magnetic field $\boldsymbol{H}$, α, β, γ

(j3) Which angular dependence would you use to discriminate between SMR and AMR, and why?

Solutions

(a) For $\alpha = \pi/2$, we have $\widehat{\boldsymbol{M}} = \widehat{\boldsymbol{y}}$, and $\boldsymbol{\mu}_s = \mu_{sy}(z)\widehat{\boldsymbol{y}}$:

$$\begin{pmatrix} \boldsymbol{J}_e \\ \boldsymbol{J}_s^y \end{pmatrix} = \sigma_N \begin{pmatrix} E_x + \dfrac{\theta_{\text{SHE}}}{2e}\dfrac{\partial \mu_{sy}}{\partial z} & 0 \\ 0 & -\theta_{\text{SHE}} E_x + \dfrac{1}{2e}\dfrac{\partial \mu_{sy}}{\partial z} \end{pmatrix} \begin{pmatrix} \widehat{\boldsymbol{x}} \\ \widehat{\boldsymbol{z}} \end{pmatrix}$$

(b) Spin accumulation diffusion equation, with $l_{sf}^* = l_{sf,N}^*$

$$\frac{d^2\mu_{sy}}{dz^2} - \frac{1}{l_{sf}^{*2}}\mu_{sy} = 0$$

For a semi-infinite layer:

$$\mu_{sy}(z) = A e^{-z/l_{sf}^*}$$

Boundary condition:

$$\boldsymbol{J}_s^y(0) = -\left[\frac{e}{2\pi} g_r^{\uparrow\downarrow} \widehat{\boldsymbol{M}} \times \left(\widehat{\boldsymbol{M}} \times \frac{\boldsymbol{\mu}_s^0}{\hbar}\right)\right] \widehat{\boldsymbol{y}}$$

Here, with $\widehat{\boldsymbol{M}} = \widehat{\boldsymbol{y}}$, and $\boldsymbol{\mu}_s = \mu_{sy}(z)\widehat{\boldsymbol{y}}$:

$$J_s^y(0) = 0$$

Hence:

$$-\sigma_N \theta_{\text{SHE}} E_x + \frac{\sigma_N}{2e} \frac{\partial \mu_{sy}}{\partial z}(0) = 0$$

$$-\theta_{\text{SHE}} E_x - \frac{A}{2el_{sf}^*} = 0$$

$$A = -2el_{sf}^* \theta_{\text{SHE}} E_x$$

$$\mu_{sy}(z) = -2el_{sf}^* \theta_{\text{SHE}} E_x e^{-z/l_{sf}^*}$$

(c) At $z = 0$, the spin current can be partially absorbed by the ferromagnet due to transfer of angular momentum *via* spin transfer torque. This is especially true for $\alpha = 0$.

(d)

$$\boldsymbol{J}_e = J_e(z)\widehat{\boldsymbol{x}}$$

$$J_e(z) = \sigma_N E_x + \theta_{\text{SHE}}^2 \sigma_N E_x e^{-z/l_{sf}^*}$$

$$J_e(z) = J_e^0 + \theta_{\text{SHE}}^2 J_e^0 e^{-z/l_{sf}^*}$$

(e) Averaging over the film thickness:

$$J_e = J_e^0 + \frac{1}{d_N} \int_0^\infty \theta_{\text{SHE}}^2 J_e^0 e^{-z/l_{sf}^*} dz$$

$$J_e = J_e^0 + \frac{l_{sf}^*}{d_N} \theta_{\text{SHE}}^2 J_e^0$$

(f) Setting $J_e = \sigma_{xx} E_x$:

$$\sigma_{xx} = \sigma_N \left(1 + \frac{l_{sf}^*}{d_N} \theta_{\mathrm{SHE}}^2\right)$$

Expanding to leading order in θ_{SHE}^2:

$$\rho_{xx} = \rho_{xx,\alpha=\pi/2} = \rho_N \left(1 - \frac{l_{sf}^*}{d_N} \theta_{\mathrm{SHE}}^2\right)$$

(g) For $\alpha = 0$, we have $\widehat{\boldsymbol{M}} = \widehat{\boldsymbol{x}}$, and $\boldsymbol{\mu}_s = \mu_{sy}(z)\widehat{\boldsymbol{y}}$:

$$\begin{pmatrix} \boldsymbol{J}_e \\ \boldsymbol{J}_s^y \end{pmatrix} = \sigma_N \begin{pmatrix} E_x + \dfrac{\theta_{\mathrm{SHE}}}{2e} \dfrac{\partial \mu_{sy}}{\partial z} & 0 \\ 0 & -\theta_{\mathrm{SHE}} E_x + \dfrac{1}{2e} \dfrac{\partial \mu_{sy}}{\partial z} \end{pmatrix} \begin{pmatrix} \widehat{\boldsymbol{x}} \\ \widehat{\boldsymbol{z}} \end{pmatrix}$$

Spin accumulation diffusion equation:

$$\frac{d^2 \mu_{sy}}{dz^2} - \frac{1}{l_{sf}^{*2}} \mu_{sy} = 0$$

For a semi-infinite layer:

$$\mu_{sy}(z) = A e^{-z/l_{sf}^*}$$

Boundary condition

$$\boldsymbol{J}_s^y(0) = -\left[\frac{e}{2\pi} g_r^{\uparrow\downarrow} \widehat{\boldsymbol{M}} \times \left(\widehat{\boldsymbol{M}} \times \frac{\boldsymbol{\mu}_s^0}{\hbar}\right)\right] \widehat{\boldsymbol{y}}$$

Here, with $\widehat{\boldsymbol{M}} = \widehat{\boldsymbol{x}}$, and $\boldsymbol{\mu}_s = \mu_{sy}(z)\widehat{\boldsymbol{y}}$:

$$J_s^y(0) = -e g_r^{\uparrow\downarrow} \mu_{sy}(0)/h$$

Hence:

$$-\sigma_N \theta_{\mathrm{SHE}} E_x + \frac{\sigma_N}{2e} \frac{\partial \mu_{sy}}{\partial z}(0) = -e g_r^{\uparrow\downarrow} \mu_{sy}(0)/h$$

$$-\sigma_N \theta_{SH} E_x - \frac{\sigma_N A}{2el^*_{sf}} = -eg_r^{\uparrow\downarrow} A/h$$

$$A = -\frac{2el^*_{sf}\theta_{\mathrm{SHE}} E_x}{1+2e^2 l^*_{sf} g_r^{\uparrow\downarrow}/(h\sigma_N)}$$

As a result:

$$\boldsymbol{J}_e = J_e(z)\hat{\boldsymbol{x}}$$

with

$$J_e(z) = \sigma_N E_x + \theta^2_{\mathrm{SHE}} \sigma_N E_x \frac{e^{-z/l^*_{sf}}}{1+2e^2 l^*_{sf} g_r^{\uparrow\downarrow}/(h\sigma_N)}$$

$$J_e(z) = J_e^0 + \theta^2_{\mathrm{SHE}} J_e^0 \frac{e^{-z/l^*_{sf}}}{1+l^*_{sf} g_r^{\uparrow\downarrow} G_0/\sigma_N}$$

with $G_0 = 2e^2/h$, the quantum conductance.
Averaging over the film thickness:

$$J_e = J_e^0 + \frac{1}{d_N}\int_0^\infty \theta^2_{\mathrm{SHE}} J_e^0 \frac{e^{-z/l^*_{sf}}}{1+l^*_{sf} g_r^{\uparrow\downarrow} G_0/\sigma_N} dz$$

$$J_e = J_e^0 + \frac{l^*_{sf}}{d_N}\theta^2_{\mathrm{SHE}} \frac{1}{1+l^*_{sf} g_r^{\uparrow\downarrow} G_0/\sigma_N} J_e^0$$

Setting $J_e = \sigma_{xx} E_x$:

$$\sigma_{xx} = \sigma_N \left(1 + \frac{l^*_{sf}}{d_N}\theta^2_{\mathrm{SHE}} \frac{1}{1+l^*_{sf} g_r^{\uparrow\downarrow} G_0/\sigma_N}\right)$$

Expanding to leading order in θ^2_{SHE}:

$$\rho_{xx} = \rho_{xx,\alpha=0} = \rho_N \left(1 - \frac{l^*_{sf}}{d_N}\theta^2_{\mathrm{SHE}} \frac{1}{1+l^*_{sf} g_r^{\uparrow\downarrow} G_0/\sigma_N}\right)$$

(h) Setting $\Delta\rho_{xx} = \rho_{xx,\alpha=0} - \rho_{xx,\alpha=\pi/2} = \rho_0 - \rho_{\pi/2}$:

$$\Delta\rho_{xx} = \rho_N\left(1 - \frac{l_{sf}^*}{d_N}\theta_{\mathrm{SHE}}^2 \frac{1}{1 + l_{sf}^* g_r^{\uparrow\downarrow} G_0/\sigma_N}\right) - \rho_N\left(1 - \frac{l_{sf}^*}{d_N}\theta_{\mathrm{SHE}}^2\right)$$

$$\Delta\rho_{xx} = \rho_N\theta_{\mathrm{SHE}}^2 \frac{l_{sf}^*}{d_N}\frac{l_{sf}^* g_r^{\uparrow\downarrow} G_0}{\sigma_N + l_{sf}^* g_r^{\uparrow\downarrow} G_0}$$

(i) Whatever γ, $\boldsymbol{M}$ is perpendicular to $\boldsymbol{\mu}_s$.

(j1) See §2.6.

(j2) Here, the angular dependence of AMR can be written as follows:

$$\rho_{xx,\mathrm{AMR}}(\alpha) = \rho_\perp - \Delta\rho_{\perp\parallel}\cos^2\alpha$$

$$\rho_{xx,\mathrm{AMR}}(\beta) = \rho_\perp$$

$$\rho_{xx,\mathrm{AMR}}(\gamma) = \rho_\perp - \Delta\rho_{\perp\parallel}\sin^2\gamma$$

with $\Delta\rho_{\perp\parallel} = \rho_\perp - \rho_\parallel$

(j3) γ can be used to discriminate between AMR and SMR. (Reminder: $\rho_{xx,\mathrm{SMR}}(\gamma) = \rho_0$)

Additional note: Here, we used a semi-infinite layer to ease the calculations. The exercise can also be done if one considers a finite size layer of thickness d_N. Basically, this change will modify the boundary condition in question (b), the upper bound of the integral in question (e), and the final results – one obtains:

$$\Delta\rho_{xx} = \rho_0\theta_{\mathrm{SHE}}^2 \frac{l_{sf}^*}{d_N}\frac{l_{sf}^* g_r^{\uparrow\downarrow} G_0 \tanh^2\frac{d_N}{2l_{sf}^*}}{\sigma_N + l_{sf}^* g_r^{\uparrow\downarrow} G_0 \coth\frac{d_N}{l_{sf}^*}} \Big/ \left(1 - \theta_{\mathrm{SHE}}^2\frac{2l_{sf}^*}{d_N}\tanh\frac{d_N}{2l_{sf}^*}\right)$$

$$\approx \rho_0\theta_{\mathrm{SHE}}^2 \frac{l_{sf}^*}{d_N}\frac{l_{sf}^* g_r^{\uparrow\downarrow} G_0 \tanh^2\frac{d_N}{2l_{sf}^*}}{\sigma_N + l_{sf}^* g_r^{\uparrow\downarrow} G_0 \coth\frac{d_N}{l_{sf}^*}}$$

The physics at stake and the main conclusions remain unchanged, in particular the θ_{SHE}^2 dependence. For a full description, see H. Nakayama *et al.*, *Phys. Rev. Lett.* **110**, 206601 (2013), and Y.-T.Chen *et al.*, *Phys. Rev. B* **87**, 144411 (2013).

6.9 Harmonic Analysis of the Anomalous Hall Voltage ($V_{xy}^{\omega}, V_{xy}^{2\omega}$) – chapters 4 and 5

■ Introduction/Reminder

This exercise illustrates the harmonic Hall voltage measurement technique, which has become a standard technique to determine the symmetry and magnitude of the effective fields H_{FL}, and H_{DL} originating the spin-orbit torques $\boldsymbol{T}_{\mathrm{FL}}$, and $\boldsymbol{T}_{\mathrm{DL}}$ (§4.7). It is based on U. H. Pi *et al.*, *Appl. Phys. Lett.* **97**, 162507 (2010).

A Hall-bar structure is patterned into a heavy metal non-magnet (N)/ferromagnet (F) bilayer and a periodic charge current $I_x = I\sin(\omega t)$ is driven along the longitudinal arm ($\hat{\boldsymbol{x}}$) of the Hall-bar, therefore generating a periodic spin-orbit torque (figure 73). In this exercise, to simplify calculations, we consider that only a periodic SOT term (§4.7) originating from a periodic effective field along $\hat{\boldsymbol{y}}$, of the form $H_{\mathrm{SOT},y} = H_{\mathrm{SOT}}\sin(\omega t)$ is generated. Under the application of a static transverse external field H_y, the tilting of the magnetization out of the easy axis (here, $\hat{\boldsymbol{z}}$) is then modulated by $H_{\mathrm{SOT},y}$ induced by I_x. The anomalous Hall resistance R_{xy}, which relates to the z-component of magnetization (chapter 5) thus takes the form:

$$R_{xy}(H) = R_{xy}(H_y) + \frac{dR_{xy}(H)}{dH} H_{\mathrm{SOT}}\sin(\omega t)$$

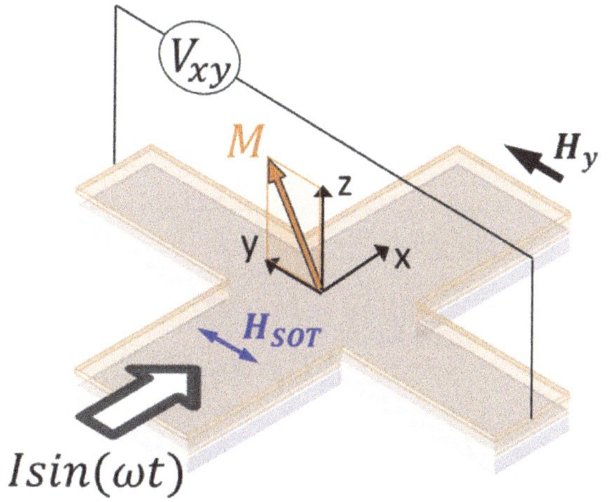

FIG. 73 – Illustration of the device used for the harmonic Hall voltage measurement technique.

■ Questions

(a) Show that the Hall voltage takes the form

$$V_{xy}(H) = V_{xy}^{0} + V_{xy}^{\omega}\sin(\omega t) + V_{xy}^{2\omega}\cos(2\omega t)$$

and explicit V_{xy}^{0}, V_{xy}^{ω}, and $V_{xy}^{2\omega}$ as a function of $R_{xy}(H_y)$, $\frac{dR_{xy}(H)}{dH}$, H_{SOT}, and I.

(b) Express H_{SOT} as a function of V_{xy}^{ω}, or its derivative, and $V_{xy}^{2\omega}$. Experimental data of V_{xy}^{ω} and $V_{xy}^{2\omega}$ are given in figure 74.
(c) Explain why $V_{xy}^{2\omega}$ is non-zero for the //Pt(3)/Co(0.6)/AlOx(1.8 nm) trilayer and zero for the //Pt(3)/Co(0.6)/Pt(3 nm) trilayer.
(d) From figure 74 and point (b), estimate and plot H_{SOT} *vs.* J_e. Comment the data.
(e) Compare the Oersted field H_{Oe} created by I to H_{SOT} and comment.

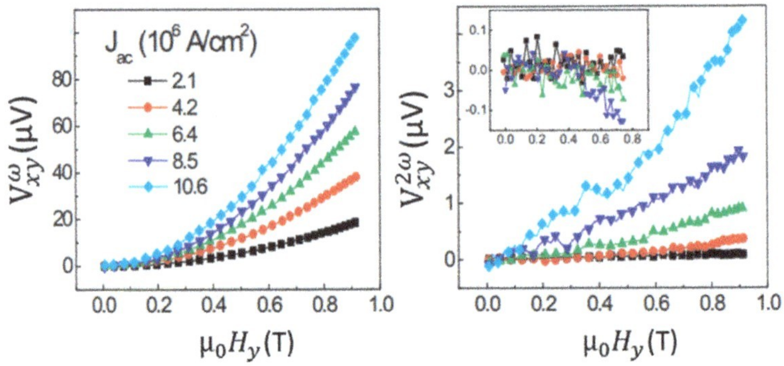

FIG. 74 – First V_{xy}^{ω} and second $V_{xy}^{2\omega}$ harmonic transverse voltages measured as a function of the external transverse field H_y for a Ta(5)//Pt(3)/Co(0.6)/AlOx(1.8 nm) trilayer (main figures) and a Ta(5)//Pt(3)/Co(0.6)/Pt(3 nm) trilayer (inset). Adapted with permission from the American Physical Society: U. H. Pi *et al.*, *Appl. Phys. Lett.* **97**, 162507 (2010). Copyright 2010.

Solutions

(a)
$$V_{xy}(H) = R_{xy}(H)I_x$$

$$V_{xy}(H) = \left(R_{xy}(H_y) + \frac{dR_{xy}(H)}{dH} H_{\text{SOT}}\sin(\omega t)\right) I\sin(\omega t)$$

$$V_{xy}(H) = V_{xy}^{0} + V_{xy}^{\omega}\sin(\omega t) + V_{xy}^{2\omega}\cos(2\omega t)$$

with

$$V_{xy}^{0} = I\frac{dR_{xy}(H)}{dH} H_{\text{SOT}}/2$$

$$V_{xy}^{\omega} = IR_{xy}(H)$$

$$V_{xy}^{2\omega} = I\frac{dR_{xy}(H)}{dH}H_{\text{SOT}}/2$$

(b)

$$H_{\text{SOT}} = 2\frac{V_{xy}^{2\omega}}{dV_{xy}^{\omega}/dH}$$

(c) $V_{xy}^{2\omega}$ is non-zero for the //Pt(3)/Co(0.6)/AlOx(1.8 nm) trilayer and zero for the //Pt(3)/Co(0.6)/Pt(3 nm) trilayer because the SOT considered and its related effective field requires symmetry breaking, here structural inversion asymmetry.

(d) Estimating and plotting H_{SOT} *vs.* J_e gives something similar to the actual data plotted below. We note that H_{SOT} is linear with J_e, as expected from its origin. Above a threshold current density $J_e \sim 8 \times 10^6$ A.cm^{-2}, Joule heating makes H_{SOT} depart from its linear relation with J_e. From the slope of H_{SOT} *vs.* J_e, we obtain $\mu_0 H_{\text{SOT}}[\text{mT}] \sim 3J_e\left[10^6\text{A.cm}^{-2}\right]$.
We note that V_{xy}^{0} is not used for the determination of H_{SOT} because the *dc* voltage is usually contaminated by spurious contributions for example due to misalignments of the different branches of the Hall bar.

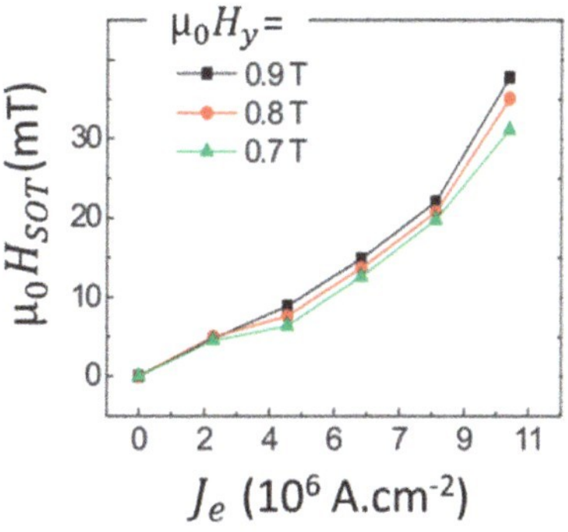

Adapted with permission from the American Physical Society.
U. H. Pi *et al.*, *Appl. Phys. Lett.* **97**, 162507 (2010). Copyright 2010.

(e) From $\mu_0 H_{Oe} * 2 * width\ \mu_0 J_e * surface$, we estimate $\mu_0 H_{Oe}[\text{mT}] \sim 0.5J_e$ $\left[10^6\text{A.cm}^{-2}\right]$. The contribution from the Oersted field is thus smaller compared to the findings.

Note: in this simplified exercise, we considered a magnetization pointing out-of-plane and a damping-like torque only, based on U. H. Pi *et al.*, *Appl. Phys. Lett.* **97**, 162507 (2010). A general scheme to measure the amplitude and the direction of the effective fields originating SOTs, when the magnetization direction is random and when field- and damping-like components are at stake, is provided in K. Garello *et al.*, *Nat. Nanotech.* **8**, 587 (2013). Symmetry arguments also allow unravelling the different contributions to the transverse voltage, including, for example, the one due to anisotropic magnetoresistance (see exercise 6.1).

6.10 Intrinsic Anomalous Hall and Nernst Effects (AHE, ANE) – chapter 5

■ Introduction/Reminder

A prototypical example illustrating Berry curvature physics is the case of a spin-polarized 2D electron gas with Rashba spin-orbit coupling (§5.4). The corresponding Hamiltonian (figure 59, bottom right) is

$$\mathcal{H} = \frac{\hbar^2 \boldsymbol{k}^2}{2m_e} + \alpha_R(\boldsymbol{k} \times \boldsymbol{\sigma}) \cdot \widehat{\boldsymbol{z}} - \Delta\boldsymbol{\sigma}_z$$

From this Hamiltonian, the aim of this exercise is to demonstrate the expressions of the eigenvalues $\varepsilon_\pm$, eigenvectors $|\pm\rangle$, and Berry curvature $\Omega^z_\pm$ [equations (107)–(109)]. The intrinsic anomalous Hall conductivity (figure 60) is then obtained from [equation (100)]

$$\sigma_{xy} = \frac{e^2}{\hbar} \int \sum_{\pm} \Omega^z_\pm(k) f(\varepsilon(k)) \frac{d^2k}{(2\pi)^2}$$

and its complementary anomalous Nernst conductivity from [equation (112)]

$$\alpha_{xy} = -2e \frac{g_0}{G_0} \frac{\partial \sigma_{xy}(\varepsilon_F)}{\partial \varepsilon}$$

■ Questions

(a) Show that the above Hamiltonian can take the form $\mathcal{H} = \dfrac{\hbar^2 \boldsymbol{k}^2}{2m_e} + \boldsymbol{b}(k) \cdot \boldsymbol{\sigma}$, with $\boldsymbol{b}(k) = (-\alpha_R k_y, \alpha_R k_x, -\Delta)$.

Reminder of the Pauli matrices: $\sigma_x = \begin{pmatrix} 0 & 1 \\ 1 & 0 \end{pmatrix}$, $\sigma_y = \begin{pmatrix} 0 & -i \\ i & 0 \end{pmatrix}$, and $\sigma_z = \begin{pmatrix} 1 & 0 \\ 0 & -1 \end{pmatrix}$.

(b) Show that the eigenvalues can be expressed as

$$\varepsilon_\pm(k) = \varepsilon_0(k) \pm |\boldsymbol{b}(k)|.$$

(c) With these eigenvalues, demonstrate that the eigenvectors of $\mathcal{H}$ are of the form

$$|\pm\rangle = \sqrt{1 \mp \frac{\Delta}{|\boldsymbol{b}(k)|}}|\psi_\Uparrow\rangle + ie^{i\phi}\sqrt{1 \pm \frac{\Delta}{|\boldsymbol{b}(k)|}}|\psi_\Downarrow\rangle$$

where $|\psi\rangle_\Uparrow = \frac{1}{\sqrt{2}}e^{i\boldsymbol{k}\cdot\boldsymbol{r}}\begin{pmatrix}1\\0\end{pmatrix}$ and $|\psi\rangle_\Downarrow = \frac{1}{\sqrt{2}}e^{i\boldsymbol{k}\cdot\boldsymbol{r}}\begin{pmatrix}0\\1\end{pmatrix}$ are the eigenvectors for free electrons.

(d) With these eigenvectors, from equation (87) $\Omega_n(\boldsymbol{k}) = \nabla_{\boldsymbol{k}} \times \langle\psi_n(\boldsymbol{k})|i\nabla_{\boldsymbol{k}}|\psi_n(\boldsymbol{k})\rangle$, show that the Berry curvature takes the form

$$\Omega_\pm^z(k) = \mp\frac{\alpha_R^2\Delta}{2\left(\alpha_R^2k^2 + \Delta^2\right)^{3/2}}$$

Use the fact that $|\pm\rangle = e^{i\boldsymbol{k}\cdot\boldsymbol{r}}|u_\pm\rangle$, and apply equation (87) to $|u_\pm\rangle$.

■ Solutions

(a) The Hamiltonian can be transformed as follows:

$$\mathcal{H} = \frac{\hbar^2\boldsymbol{k}^2}{2m_e} + \alpha_R(\boldsymbol{k}\times\boldsymbol{\sigma})\cdot\hat{\boldsymbol{z}} - \Delta\boldsymbol{\sigma}_z$$

$$\mathcal{H} = \frac{\hbar^2\boldsymbol{k}^2}{2m_e} + \alpha_R(\hat{\boldsymbol{z}}\times\boldsymbol{k})\cdot\boldsymbol{\sigma} - \Delta\boldsymbol{\sigma}_z$$

$$\mathcal{H} = \frac{\hbar^2\boldsymbol{k}^2}{2m_e} + \boldsymbol{b}(k)\cdot\boldsymbol{\sigma}$$

with $\boldsymbol{b}(k) = \left(-\alpha_Rk_y, \alpha_Rk_x, -\Delta\right)$.

(b) In this type of formulation, the eigenvalues can be expressed as

$$\varepsilon_\pm(k) = \varepsilon_0(k) \pm |\boldsymbol{b}(k)|$$

$$\varepsilon_\pm(k) = \frac{\hbar^2k^2}{2m_e} \pm \sqrt{\alpha_R^2k^2 + \Delta^2}$$

More precisely, in the formulation of the Hamiltonian, $\boldsymbol{b}(k)$ represents a point on the Bloch sphere, which has the consequence that the solutions of the eigenvectors of the total Hamiltonian are in the $\left(|\psi_{\Uparrow}\rangle, |\psi_{\Downarrow}\rangle\right)$ basis of the original wavefunctions in the absence of the $\boldsymbol{b}(k)\cdot\boldsymbol{\sigma}$ perturbation. Thus, the Schrödinger equation

$$\mathcal{H}|\pm\rangle = \varepsilon|\pm\rangle$$

has eigenvectors of the form

$$|\pm\rangle = A_{\pm}|\psi_{\Uparrow}\rangle + B_{\pm}|\psi_{\Downarrow}\rangle$$

with $|\psi_{\Uparrow}\rangle = \frac{1}{\sqrt{2}} e^{i\boldsymbol{k}\cdot\boldsymbol{r}}\begin{pmatrix}1\\0\end{pmatrix}$ and $|\psi_{\Downarrow}\rangle = \frac{1}{\sqrt{2}} e^{i\boldsymbol{k}\cdot\boldsymbol{r}}\begin{pmatrix}0\\1\end{pmatrix}$, the eigenvectors for free electrons.

After insertion into the Schrödinger equation, and taking $k_x + ik_y = k(\cos\phi + i\sin\phi) = ke^{i\phi}$, and thus $k_x - ik_y = k(\cos\phi - i\sin\phi) = ke^{-i\phi}$, we get

$$\begin{pmatrix} \dfrac{\hbar^2k^2}{2m_e} - \Delta - \varepsilon & -i\alpha_R k e^{-i\phi} \\ i\alpha_R k e^{i\phi} & \dfrac{\hbar^2k^2}{2m_e} + \Delta - \varepsilon \end{pmatrix}\begin{pmatrix} A_{\pm} \\ B_{\pm}\end{pmatrix} = 0$$

Zeroing the determinant of the square matrix on the left returns the energy eigenvalues

$$\varepsilon_{\pm}(k) = \frac{\hbar^2k^2}{2m_e} \pm \sqrt{\alpha_R^2k^2 + \Delta^2} = \varepsilon_0(k) \pm |\boldsymbol{b}(k)|$$

(c) We now detail how to find the coefficients A_+ and B_+, which govern the eigenvector $|+\rangle = A_+|\psi_{\Uparrow}\rangle + B_+|\psi_{\Downarrow}\rangle$ associated to the eigenvalue ε_+ Substituting ε_+ in the above matrix, we get

$$\begin{pmatrix} -\Delta - |\boldsymbol{b}| & -i\alpha_R k e^{-i\phi} \\ i\alpha_R k e^{i\phi} & \Delta - |\boldsymbol{b}| \end{pmatrix}\begin{pmatrix} A_+ \\ B_+ \end{pmatrix} = 0$$

This leads to a system of 2 equations and 2 unknowns

$$(-\Delta - |\boldsymbol{b}|)A_+ - i\alpha_R k e^{i\phi} B_+ = 0$$

$$i\alpha_R k e^{-i\phi} A_+ + (\Delta - |\boldsymbol{b}|)B_+ = 0$$

However, the system is degenerate, so we need to choose one variable and find the other. Starting from zero, we would choose $A_+ = 1$, then find B_+, then use $A_+^* A_+ + B_+^* B_+ = 1$ for normalization. Because the solution is given to us, we choose $A_+ = \sqrt{1 - \frac{\Delta}{|\boldsymbol{b}|}}$ and find B_+ from

$$B_+ = -\frac{\Delta + |\boldsymbol{b}|}{i\alpha_R k e^{i\phi}} A_+ = -\frac{\Delta + |\boldsymbol{b}|}{i\alpha_R k e^{i\phi}} \sqrt{1 - \frac{\Delta}{|\boldsymbol{b}|}}$$

$$B_+ = -\frac{i\alpha_R k e^{-i\phi}}{\Delta - |\boldsymbol{b}|} A_+ = -\frac{i\alpha_R k e^{-i\phi}}{\Delta - |\boldsymbol{b}|} \sqrt{1 - \frac{\Delta}{|\boldsymbol{b}|}}$$

$$B_+^2 = e^{-2i\phi} \frac{\Delta + |\boldsymbol{b}|}{\Delta - |\boldsymbol{b}|} \left(1 - \frac{\Delta}{|\boldsymbol{b}|}\right) = i^2 e^{-2i\phi} \left(1 + \frac{\Delta}{|\boldsymbol{b}|}\right)$$

$$B_+ = ie^{-i\phi} \sqrt{1 + \frac{\Delta}{|\boldsymbol{b}|}}$$

The coefficients A_- and B_-, which govern the eigenvector $|-\rangle = A_-|\psi_\Uparrow\rangle + B_-|\psi_\Downarrow\rangle$ associated to the eigenvalue ε_- can be found by using the same methodology.
From

$$|\pm\rangle = A_\pm|\psi_\Uparrow\rangle + B_\pm|\psi_\Downarrow\rangle$$

with $|\psi_\Uparrow\rangle = \frac{1}{\sqrt{2}} e^{i\boldsymbol{k}\cdot\boldsymbol{r}} \begin{pmatrix} 1 \\ 0 \end{pmatrix}$ and $|\psi_\Downarrow\rangle = \frac{1}{\sqrt{2}} e^{i\boldsymbol{k}\cdot\boldsymbol{r}} \begin{pmatrix} 0 \\ 1 \end{pmatrix}$, and having determined A_+, B_+, A_-, and B_-, we get the expressions of the eigenvectors

$$|\pm\rangle = \sqrt{1 \mp \frac{\Delta}{|\boldsymbol{b}(k)|}} |\psi_\Uparrow\rangle + ie^{i\phi} \sqrt{1 \pm \frac{\Delta}{|\boldsymbol{b}(k)|}} |\psi_\Downarrow\rangle$$

(d) The expressions of the eigenvectors can be transformed into

$$|\pm\rangle = e^{i\boldsymbol{k}\cdot\boldsymbol{r}} |u_\pm\rangle$$

with

$$|u_{\pm}\rangle = \frac{1}{\sqrt{2}}\sqrt{1 \mp \frac{\Delta}{\left(\alpha_R^2 k^2 + \Delta^2\right)^{1/2}}}\begin{pmatrix}1\\0\end{pmatrix} + i\frac{1}{\sqrt{2}}e^{i\phi}\sqrt{1 \pm \frac{\Delta}{\left(\alpha_R^2 k^2 + \Delta^2\right)^{1/2}}}\begin{pmatrix}0\\1\end{pmatrix}$$

and $\phi = \arctan\left(k_y/k_x\right)$ and $k^2 = k_x^2 + k_y^2$.
We now detail the steps necessary to calculate $\mathbf{\Omega}_+^z(k)$, which is the Berry curvature along z corresponding to the eigenvector $|u_+\rangle$.
Take

$$|u_+\rangle = \begin{pmatrix} f_1\left(k_x, k_y\right) \\ f_2\left(k_x, k_y\right) \end{pmatrix}$$

$$\langle u_+| = \begin{pmatrix} f_1^*\left(k_x, k_y\right) & f_2^*\left(k_x, k_y\right) \end{pmatrix}$$

We then have

$$\langle u_+|\nabla_{\boldsymbol{k}}|u_+\rangle = \begin{pmatrix} f_1^*\dfrac{\partial f_1}{\partial k_x} + f_2^*\dfrac{\partial f_2}{\partial k_x} \\ f_1^*\dfrac{\partial f_1}{\partial k_y} + f_2^*\dfrac{\partial f_2}{\partial k_y} \\ 0 \end{pmatrix} = \begin{pmatrix} Q_x \\ Q_y \\ 0 \end{pmatrix}$$

Thus, we get

$$\Omega_n^z(k) = \mathbf{\Omega}_n(\boldsymbol{k}) \cdot \hat{\boldsymbol{z}}$$

$$\Omega_n^z(k) = \left(\nabla_{\boldsymbol{k}} \times \langle u_+|i\nabla_{\boldsymbol{k}}|u_+\rangle\right) \cdot \hat{\boldsymbol{z}}$$

$$\Omega_n^z(k) = -\Im m\left[\left(\frac{\partial Q_y}{\partial k_x} - \frac{\partial Q_x}{\partial k_y}\right)\right] \cdot \hat{\boldsymbol{z}}$$

$$\mathbf{\Omega}_+^z(k) = -\frac{\alpha_R^2 \Delta}{2\left(\alpha_R^2 k^2 + \Delta^2\right)^{3/2}}$$

Using the same methodology for $\mathbf{\Omega}_-^z(k)$, we obtain

$$\mathbf{\Omega}_\pm^z(k) = \mp\frac{\alpha_R^2 \Delta}{2\left(\alpha_R^2 k^2 + \Delta^2\right)^{3/2}}$$

Note 1: the eigenvalues, eigenvectors, and (zero) Berry curvature for a Rashba Hamiltonian (non spin-split) can be straightforwardly obtained from the above expressions by setting $\Delta = 0$ (§5.4, figure 59 top right).

Note 2: to go further, the general expressions of eigenvalues, and Berry curvature can be found in D. Culcer *et al.*, *Phys. Rev. B* **68**, 045327 (2003) for two-level degenerate systems with a Hamiltonian of the generic form $\mathcal{H} = \frac{\hbar^2 \boldsymbol{k}^2}{2m_e} + \boldsymbol{b}(k) \cdot \boldsymbol{\sigma}$ with $\boldsymbol{b}(k) = (-\alpha_R f(k), \alpha_R f(k), -\Delta)$. In particular, it is possible to show that $\boldsymbol{\Omega} \propto \frac{\boldsymbol{b}}{|\boldsymbol{b}|^3}$, which highlights the expression of a 'magnetic monopole' in k-space, sitting at $\boldsymbol{b} = 0$ [D. Xiao *et al.*, *Rev. Mod. Phys.* **82**, 1959 (2010)].

Chapter 7

Conclusion

By now, the objectives of giving the reader the tools to go further and understand other related or more elaborate phenomena of spintronics should have been achieved. Of course, "Everybody calls 'clear' those ideas which have the same degree of confusion as his own" – M. Proust. The aim of this conclusion is to open up the field of vision by presenting in a non-exhaustive way some of the current topics to which the interested reader may wish to turn. To this end, the selected topics are briefly mentioned and accompanied by references consisting of comprehensive review articles that provide a complete overview of the topic.

It is important to mention at this point that the experimental measurement of the effects presented in this book is extremely delicate as it involves the fabrication of nano-sized structures, the detection of signals that can be of a few nanovolts and that can be obtained in some cases by means of pulses of a few nanoseconds or even a few picoseconds, as well as the use of advanced electrical, optical, and thermal techniques, not to mention the use of large instruments such as synchrotrons.

In general, spintronics and the disciplines that interface with it, such as microelectronics, quantum computing, biology, and medicine, call for teams of researchers and engineers with a wide range of skills in, for example:

- **solid state physics**
- **materials science and engineering**
- **instrumentation**
- **theory**
- **analytical and numerical computation**
- **signal processing**
- **integrated circuit design.**

This wealth of skills needed by the discipline and of interest to society promises a bright future for those who wish to join it, whether they approach the discipline from the point of view of fundamental science or from an applicative perspective, in physics, mathematics, biology, or chemistry.

DOI: 10.1051/978-2-7598-2917-0.c007

Moreover, the great diversity in which magnetic materials appear in nature – from metals to semimetals to insulators, not to mention the great wealth of magnetic textures and topological structures described by group theory, provides a vast and fascinating playground. Among the materials that are currently the subject of many studies are, for example:

- **oxides** with their unique versatility and multiple functionalities – M. Coll *et al.*, *Appl. Surf. Sci.* **482**, 1 (2019)
- **2D materials** because of their unusual nanoscale-related characteristics – S. Z. Butler *et al.*, *ACS Nano* **7**, 2898 (2013), Q. H. Wang *et al.*, *ACS Nano* **16**, 6960 (2022)
- other topological materials and textures, including:
    - **topological insulators** exhibiting topologically protected spin-polarized surface states – B. A. Bernevig *et al.*, *Nature* **603**, 41 (2022)
    - **Weyl semimetals** displaying Dirac nodes of opposite chirality – N. Kumar *et al.*, *Chem. Rev.* **121**, 2780 (2021)
    - the newly discovered **unconventional 'alter' magnets** with anisotropic higher-order partial-wave forms of magnetically ordered phases – L. Smejkal *et al.*, *Nat. Rev. Mater.* **7**, 482 (2022)
    - **topological magnetic textures** like skyrmions and other non-collinear structures, exhibiting locally topologically protected states as opposed to the surrounding magnetic structure – A. Soumyanarayanan *et al.*, *Nature* **539**, 509 (2016), B. Göbel *et al.*, *Phys. Rep.* **895**, 1 (2021)
- **antiferromagnets** for which the presence of two sub-lattices of magnetic moments gives them unique properties in real and reciprocal space – T. Jungwirth *et al.*, *Nat. Nano.* **11**, 231 (2016), V. Baltz *et al.*, *Rev. Mod. Phys.* **90**, 015005 (2018)
- **ferrimagnets**, which combine at compensation some favorable properties of an antiferromagnet with the net spin polarization of a ferromagnet – S. K. Kim *et al.*, *Nat. Mater.* **21**, 24 (2022)
- **heusler alloys** for their specific density of states and corresponding large spin polarization and low damping – T. Graf *et al.*, *Prog. Solid State Chem.* **39**, 1 (2011), K. Elphick *et al.*, *Sci. Tech. Adv. Mater.* **22**, 235 (2021)
- **organic materials**, which are more abundant and cheaper than other materials and could be used for molecular spintronics – A. Cornia *et al.*, *Nat. Mater.* **16**, 505 (2017)
- **artificial spin structures**, such as artificial spin ices, which give access to some non-conventional behaviors such as emergent magnetic monopoles and their dynamics – S. H. Skjærvø *et al.*, *Nat. Rev. Phys.* **2**, 13 (2020)
- **heterostructures** combining different spin orders or ways to propagate angular momentum, thus being a source of new physics and a promise of new functionalities that arise from proximity effects and competing interactions. Striking and

exciting examples of how several areas of condensed matter physics can look forward to a common future are provided by the combination of:

- **magnetic and superconducting order**, the latter bringing quantum coherence and dissipationless transport – J. Linder *et al.*, *Nat. Phys.* **11**, 307 (2015)
- **strongly correlated antiferromagnetism and high-T_c superconductivity**, where the intermingling between spin transport and antiferromagnetic order proves to be decisive – F. Giustino *et al.*, *J. Phys. Mater.* **3**, 042006 (2020)

The wealth of materials available, both crystalline and amorphous, allows the selection of specific materials and combinations of magnetic (*e.g.*, anisotropy, exchange, DMI, damping) and transport (*e.g.*, polarization, conductivity, Berry curvature) properties for the study of specific current research areas, whether for example:

- **symmetries and topology**, towards quantum (band-topology) spintronics and non-dissipative spin-conservative electronics – C. K. Chiu *et al.*, *Rev. Mod. Phys.* **88**, 035005 (2016)
- **angular momentum transport and transfer**, where several carriers of angular momentum are involved – electron, phonon, and photon – and wherein the amplitude or the phase is used for information encoding and transfer. This research includes several fields:
  - **spin currents/spin orbitronics** – S. Maekawa *et al.* (eds), *Spin Currents*, Oxford University Press (2012)
  - **orbital spintronics** – D. Go *et al.*, *Europhys. Lett.* **135**, 37001 (2021)
  - **spin mechatronics** – M. Matsuo *et al.*, *J. Phys. Soc. Jpn* **86**, 011011 (2017), J. Puebla *et al.*, *Appl. Phys. Lett.* **120**, 220502 (2022)
  - **opto spintronics** – A. Kirilyuk *et al.*, *Rev. Mod. Phys.* **82**, 2731 (2010)
  - **magnonics** – A. V. Chumak *et al.*, *Nat. Phys.* **11**, 453 (2015), *IEEE Trans. Mag.* **58**, 0800172 (2022)
- **spin caloritronics**, which considers the interaction between heat and spin, for example, for thermal energy conversion – S. R. Boona *et al.*, *Energy Environ. Sci.* **7**, 885 (2014), K.-i. Uchida *et al.*, *J. Phys. Soc. Jpn* **90**, 122001 (2021)
- **magneto-ionic and electrostatic gating**, in which an excess/depletion of charges is used to modify magnetism – M. Nichterwitz, *Appl. Phys. Lett. Mater.* **9**, 030903 (2021)
- **ultimate time scales**, at which specific ultrafast physical processes occur like heat to spin conversion through lattice excitation, with potential for THz applications – J. Walowski *et al.*, *J. Appl. Phys.* **120**, 140901 (2016)
- **magnetic nanoparticles**, which are highly-functionalizable objects for potential use in the health sector – K. Mahmoudi *et al.*, *Int. J. Hyperthermia* **34**, 1316 (2018), C. Naud *et al.*, *Nanoscale Adv.* **2**, 3632 (2020)

- **3D spintronics**, wherein complex magnetic configurations, like Bloch points, with specific properties become possible – A. Fernández-Pacheco, *Nat. Commun.* **8**, 15756 (2017)
- and more mature fields at present, like:
  - **MRAM/design** for IT applications like memory, processors, data security – B. Dieny *et al.* (eds), *Introduction to Magnetic Random-Access Memory*, Wiley-IEEE Press (2016), V. Krizakova *et al.*, *J. Magn. Magn. Mater.* **562**, 169692 (2022)
  - **sensors** for IT, automotive, space, and biological applications – B. Dieny *et al.*, *Nat. Elec.* **3**, 446 (2020)
  - **oscillators** for use as transceivers for telecommunication applications or neurons and axons for neuromorphic computing and related AI applications – J. Grollier, *Nat. Elec.* **3**, 360 (2020)

Index of Key Concepts

Abbreviations, Symbols and Units

1. Abbreviations

AHE	anomalous Hall effect
AMR	anisotropic magnetoresistance
AP	antiparallel
CIP	current in plane
CPP	current perpendicular-to-plane
crit	critical
DL	damping-like
DOS	density of states
DW	domain wall
DWMR	domain wall magnetoresistance
eff	effective
e.g.	*exemplī grātiā* (for example)
et al.	*et alii* (and others)
F	ferromagnet
FL	field-like
GMR	giant magnetoresistance
ibid.	*ibidem* (in the same place)
i.e.	*id est* (that is)
IEC	interlayer exchange coupling
ISHE	inverse spin Hall effect
LLG	Landau–Lifshitz–Gilbert
N	non-magnet
OHE	ordinary Hall effect
P	parallel
QHE	quantum Hall effect
QWS	quantum well state
REE	Rashba–Edelstein effect
RKKY	Ruderman–Kittel–Kasuya–Yosida
SAHE	spontaneous anomalous Hall effect
SHE	spin Hall effect
SML	spin memory loss
SO	spin-orbit
SOT	spin-orbit torque
SP	spin pumping
STT	spin transfer torque
THE	topological Hall effect
TMR	tunnel magnetoresistance

2. Symbols and Units

Bold is used for vectors, *e.g.* $\boldsymbol{r} = \overrightarrow{r}$, an *overcaret* is used for unit vectors, *e.g.* $\widehat{\boldsymbol{r}} = \boldsymbol{r}/r$, an *overdot* indicates a time-derivative, *e.g.* $\dot{\boldsymbol{r}} = \partial\boldsymbol{r}/\partial t$, a double bar is used for second-order tensor, *e.g.* $\overline{\overline{J}}_s = \left[\boldsymbol{J}_s^{(x)} \quad \boldsymbol{J}_s^{(y)} \quad \boldsymbol{J}_s^{(z)}\right] = \begin{bmatrix} J_s^{xx} & J_s^{yx} & J_s^{zx} \\ J_s^{xy} & J_s^{yy} & J_s^{zy} \\ J_s^{xz} & J_s^{yz} & J_s^{zz} \end{bmatrix}$.

Many of the parameters used are spin-dependent. In the text, they are accompanied by arrows as superscripts, where necessary. Additionally, some of the abbreviations detailed above are combined as subscripts or superscripts with the symbols, where necessary to specify that a given effect relates to a given phenomenon, which is then detailed in the text.

Some symbols may play a dual role, in which case their meaning is obvious from the context.

Symbol	Formula	Quantity	Unit	Value
Physical constants				
c		velocity of light	$\mathrm{m\ s^{-1}}$	2.998×10^{8}
μ_0		magnetic permeability of free space	$\mathrm{T\ m\ A^{-1}}$	$4\pi \times 10^{-7}$
k_B		Boltzmann constant	$\mathrm{J\ K^{-1}}$	1.381×10^{-23}
e		elementary charge	C	1.602×10^{-19}
m_e		electron mass	kg	9.109×10^{-31}
h		Planck constant	J s	6.626×10^{-34}
$\hbar$	$\frac{h}{2\pi}$	Planck constant/(2π)	J s	1.055×10^{-34}
μ_B	$\frac{e\hbar}{2m_e}$	Bohr magneton	$\mathrm{A\ m^2}$	9.274×10^{-24}
G_0	$\frac{2e^2}{h}$	electrical conductance quantum	Ω^{-1} (S)	7.748×10^{-5}
R_K	$\frac{2}{G_0}$	von Klitzing constant	Ω	2.581×10^{4}
g_0	$\frac{\pi^2 k_\mathrm{B}^2 T}{3h}$	thermal conductance quantum	$\mathrm{W\ K^{-2}}$	$9.464 \times 10^{-13} \times T$
Φ_0	$\frac{h}{2e}$	magnetic flux quantum	$\mathrm{T\ m^2}$ (Wb)	2.068×10^{-15}
K_J	$\frac{1}{\Phi_0}$	Josephson constant	$\mathrm{T^{-1}\ m^{-2}}$ ($\mathrm{Wb^{-1}}$)	4.836×10^{14}

Symbol	Formula	Quantity	Unit
Arrows and relational operators			
$\uparrow$		(spin) magnetic moment $\boldsymbol{m}$ orientation[1]	
$\downarrow$		(spin) magnetic moment $\boldsymbol{m}$ orientation[1]	
$\Uparrow$		spin angular momentum $\boldsymbol{s}$ orientation[1]	
$\Downarrow$		spin angular momentum $\boldsymbol{s}$ orientation[1]	
$\uparrow\downarrow$		spin mixing = moment mixing	
$\parallel$		parallel	
$\perp$		perpendicular	
Operators, distribution functions and related quantities			
$\boldsymbol{\nabla}$	$\left(\frac{\partial}{\partial x}, \frac{\partial}{\partial y}, \frac{\partial}{\partial z}\right)^{\mathrm{T}}$	gradient formula in cartesian coordinates	
$\boldsymbol{\nabla}\cdot$	$\frac{\partial}{\partial x} + \frac{\partial}{\partial y} + \frac{\partial}{\partial z}$	divergence formula in cartesian coordinates	
T		transpose	
$\mathcal{H}$		Hamiltonian	
$\boldsymbol{\sigma}$	$\sigma^x = \begin{pmatrix} 0 & 1 \\ 1 & 0 \end{pmatrix}$ $\sigma^y = \begin{pmatrix} 0 & -i \\ i & 0 \end{pmatrix}$ $\sigma^z = \begin{pmatrix} 1 & 0 \\ 0 & -1 \end{pmatrix}$	Pauli operator	
$\mathbf{S}$	$\frac{\hbar}{2}\boldsymbol{\sigma}$	spin angular momentum operator	
$\mathbf{L}$	$i\hbar\left(\frac{\widehat{\boldsymbol{\theta}}}{\sin\theta}\frac{\partial}{\partial\phi} - \widehat{\boldsymbol{\phi}}\frac{\partial}{\partial\theta}\right)$	orbital angular momentum operator formula in spherical coordinates	
$\mathbf{S}^{+(-)}$		spin angular momentum ladder operator	
$\mathbf{L}^{+(-)}$		orbital angular momentum ladder operator	
$\mathbf{v}$	$-i\frac{\hbar}{m_e}\boldsymbol{\nabla}$	velocity operator	
ψ		wavefunction	$\mathrm{m}^{-3/2}$
$\mathcal{P}$		parity or space inversion symmetry	
$\mathcal{T}$		time reversal symmetry	

(continued).

Symbol	Formula	Quantity	Unit
f	$\dfrac{1}{\exp\left(\frac{\varepsilon-\mu_n}{k_B T}\right)+1}$	Fermi–Dirac distribution function	
g	$f+\delta g$	distribution function	
δg		out-of-equilibrium part of g	
Quantum numbers			
n		principal quantum number	
s		spin quantum number	
m_s		spin projection quantum number	
l		orbital quantum number	
m_l		orbital projection quantum number	
j		total angular momentum quantum number	
m_j		total angular momentum projection quantum number	
Energies, *forces* and related quantities			
ε		energy	J
ε_n		energy of the nth band	J
ε_F		Fermi energy	J
Δ		Zeeman energy	J
ξ_{SO}		spin-orbit coupling energy	J
J_{sd}		exchange constant	J
μ_e	$-e\phi$	electrostatic potential	J
μ_n	$\mu_{n_0}+\dfrac{\delta n}{N(\varepsilon_F)}$	chemical potential	J
μ_{n_0}	ε_F at $T=0$	equilibrium chemical potential	J
μ	$\mu_e+\mu_n$	electrochemical potential	J
$\boldsymbol{\mu}_s$	$-(\boldsymbol{\mu}^{\uparrow}-\boldsymbol{\mu}^{\downarrow})$	spin $\boldsymbol{s}$ accumulation$^{(2)}$	J
ϕ		electric potential	V
$\boldsymbol{E}$	$-\boldsymbol{\nabla}\phi$	electric field	V m^{-1}
$\boldsymbol{F}$		force	N
α_R		Rashba parameter	J m

(continued).

Symbol	Formula	Quantity	Unit
Times and related quantities			
t		time	s
τ_e		electronic relaxation time	s
τ_{sd}	$\hbar/J_{sd}$	exchange relaxation time	s
τ_{sf}		spin diffusion relaxation time	s
ξ	$\frac{\tau_{sd}}{\tau_{sf}}$	spin mistracking parameter	
f		frequency	s^{-1}
ω	$2\pi f$	angular frequency	rad s^{-1}
Lengths, areas, volumes, angles and related quantities			
d		thickness	m
A		area	m^2
V		volume	m^3
x, y, z		cartesian coordinates	m,m,m
r, θ, ϕ		spherical coordinates	m,rad, rad
k		wavevector	m^{-1}
k_F		Fermi wavevector	m^{-1}
$k_{F,es}$		extremal spanning wavevector of the Fermi surface	m^{-1}
λ_e	$v_F\tau_e$	electronic mean free path	m
l_{sd}	$\sqrt{D\tau_{sd}}$	exchange length	m
λ_{sf}	$v_F\tau_{sf}$	spin-flip length	m
l_{sf}	$\sqrt{D\tau_{sf}}$	spin-diffusion length	m
l^*_{sf}	$\left(\frac{1}{l^{\uparrow\,2}_{sf}}+\frac{1}{l^{\downarrow\,2}_{sf}}\right)^{-1/2}$	*average* spin-diffusion length(3)	m
w_{DW}		domain wall width	m
$\boldsymbol{\Omega}_n$	$\nabla_{\boldsymbol{k}} \times \langle\psi_n(\boldsymbol{k})\|i\nabla_{\boldsymbol{k}}\|\psi_n(\boldsymbol{k})\rangle$	Berry curvature of the nth band	m^2
$\mathcal{A}_n$	$\langle\psi_n(\boldsymbol{k})\|i\nabla_{\boldsymbol{k}}\|\psi_n(\boldsymbol{k})\rangle$	Berry connection of the nth band	m
γ_n	$\oint \mathcal{A}_n(k)\,dk = \iint \Omega_n(\boldsymbol{k})\,d^2k$	Berry phase of the nth band	rad
C_n	$\frac{1}{2\pi}\oiint \Omega_n(\boldsymbol{k})\,d^2k$	Chern number of the nth band, integer	

(continued).

Symbol	Formula	Quantity	Unit
Particle density and density of states			
n	$n_0 + \delta n$	electron density	m^{-3}
n_0	$\int_{-\infty}^{\mu_{n_0}} f(\varepsilon) N(\varepsilon)\, d\varepsilon$	equilibrium electron density	m^{-3}
δn		excess electron density	m^{-3}
N		density of states	$J^{-1}\ m^{-3}$
ρ	$-e\langle\psi\|\psi\rangle$	charge density	$C\ m^{-3}$
Velocities, diffusion			
v		electron velocity	$m\ s^{-1}$
v_F		Fermi velocity	$m\ s^{-1}$
D	$\frac{1}{3} v_F \lambda_e$	electronic diffusion coefficient	$m^2\ s^{-1}$
Conductivities, resistivities and related quantities			
σ	$N(\varepsilon_F) e^2 D$	electric conductivity	$S\ m^{-1}$
σ^*	$\sigma(1-\beta^2)$	effective electric conductivity	$S\ m^{-1}$
ρ	$\frac{1}{\sigma}$	electric resistivity	$\Omega\ m$
ρ^*	$\frac{1}{\sigma^*}$	effective electric resistivity	$\Omega\ m$
ρl_{sf}^*		spin 'resistance' (resistance area product)	$\Omega\ m^2$
σ / l_{sf}^*		spin 'conductance' (conductance per unit area)(4)	$S\ m^{-2}$
r_s	$\frac{-\Delta\mu_e}{eJ_e}$	spin-coupled interface 'resistance' (resistance area product)	$\Omega\ m^2$
α	$\frac{\rho^{\downarrow}}{\rho^{\uparrow}}$	bulk spin asymmetry parameter	
γ		interface spin asymmetry parameter	
δ		interface spin-flip parameter	
β	$\frac{\alpha-1}{\alpha+1}$	spin polarization	
$G^{\uparrow\downarrow}$	$\frac{e^2}{h}\sum_{n\in N}\left[1-\sum_{m\in N} r_{\uparrow}^{nm}\left(r_{\downarrow}^{nm}\right)^*\right]$	spin mixing conductance(5),(6)	S

(continued).

Symbol	Formula	Quantity	Unit
$g^{\uparrow\downarrow}$	$\frac{G^{\uparrow\downarrow}}{AG_0/2}$	spin mixing conductance per unit area A per quantum conductance per spin channel $G_0/2$[7]	m^{-2}
$G_r^{\uparrow\downarrow}$	$\mathcal{R}e(G^{\uparrow\downarrow})$	real part of $G^{\uparrow\downarrow}$	S
$G_i^{\uparrow\downarrow}$	$\Im m(G^{\uparrow\downarrow})$	imaginary part of $G^{\uparrow\downarrow}$	S
$g_{r,\text{eff}}^{\uparrow\downarrow}$	$\left[\frac{1}{g_r^{\uparrow\downarrow}} + \frac{2l_{sf}^*}{G_0\sigma_N}\tanh^{-1}\left(\frac{d_N}{l_{sf}^*}\right)\right]^{-1}$	effective spin mixing conductance per unit area per quantum conductance per spin channel	m^{-2}
$\mathscr{g}^{\uparrow\downarrow}$	$1-\left(t_\uparrow t_\downarrow^* + r_\uparrow r_\downarrow^*\right)$	spin mixing coefficient[6]	
$\boldsymbol{h}$	$\left(\sigma_{zy}, \sigma_{xz}, \sigma_{yx}\right)^{\mathrm{T}}$	Hall vector	$(S\ m^{-1})$
Charge and spin current densities and related quantities			
$\boldsymbol{j}$	$\sigma\frac{\boldsymbol{\nabla}\mu}{e}$	partial charge current density	$A\ m^{-2}$
$\boldsymbol{J}_e$	$\boldsymbol{j}^\uparrow + \boldsymbol{j}^\downarrow$	total charge current density	$A\ m^{-2}$
$\boldsymbol{J}_s$	$-(\boldsymbol{j}^\uparrow - \boldsymbol{j}^\downarrow)$	total spin $\boldsymbol{s}$ current density[8]	$A\ m^{-2}$
$\boldsymbol{Q}$	$\frac{\hbar}{2e}\boldsymbol{J}_s = \mathcal{R}e(\langle\psi\vert\mathbf{S}\otimes\mathbf{v}\vert\psi\rangle)$	total spin $\boldsymbol{s}$ current density[8]	$J\ m^{-2}$
$\boldsymbol{J}_H$	$\boldsymbol{h}\times\boldsymbol{E}$	total charge Hall current density	$A\ m^{-2}$
θ_{SHE}	$\frac{J_e}{J_s}$	spin Hall angle	
Magnetizations and flux densities			
$\boldsymbol{M}$	$\mu_B(\boldsymbol{n}^\uparrow - \boldsymbol{n}^\downarrow)$	magnetization	$A\ m^{-1}$
M_s		saturation magnetization	$A\ m^{-1}$
$\boldsymbol{m}$	$-\vert\gamma\vert\boldsymbol{s}$	magnetic moment	$A\ m^2$
$\boldsymbol{m}$	$\boldsymbol{m}/V$	magnetic moment density[9]	$A\ m^{-1}$
$\mu_0\boldsymbol{H}$		magnetic flux density or magnetic B-field	T
$\mu_0\boldsymbol{H}_{\text{eff}}$		effective magnetic flux density	T
$\mu_0 h_{rf}$		radiofrequency magnetic flux density	T
$\mu_0\boldsymbol{H}_{SO}$	$-\frac{(\boldsymbol{v}\times\boldsymbol{E})}{c^2}$	spin-orbit magnetic flux density	T
τ_{DL}	$-\boldsymbol{T}_{\text{DL}} = \frac{\tau_{\text{DL}}}{M_S^2}\boldsymbol{M}\times(\boldsymbol{m}\times\boldsymbol{M})$	damping-like magnetic flux density amplitude	T
τ_{FL}	$-\boldsymbol{T}_{\text{FL}} = \frac{\tau_{\text{FL}}}{M_S}\boldsymbol{m}\times\boldsymbol{M}$	field-like magnetic flux density amplitude	T

(continued).

Symbol	Formula	Quantity	Unit
Angular momenta, torques and related quantities			
$\boldsymbol{s}$	$-\frac{1}{\lvert\gamma\rvert}\boldsymbol{m}$	spin angular momentum	J s
s	$\langle\psi\lvert\mathbf{S}\rvert\psi\rangle = \boldsymbol{s}/V$	spin angular momentum density	J s m^{-3}
$\boldsymbol{S}_M$	$-\frac{1}{\lvert\gamma\rvert}\boldsymbol{M}$	(macro)spin angular momentum density of a ferromagnet	J s m^{-3}
ℓ		orbital angular momentum	J s
$\boldsymbol{T}_V$	$V\boldsymbol{T}$	torque from a spin density on a magnetic moment	J
$\boldsymbol{T}$	$\propto -\frac{1}{\lvert\gamma\rvert}\frac{d\boldsymbol{M}}{dt}$	torque from a spin density on a magnetization[10]	J m^{-3}
t		transmission coefficient	
r		reflection coefficient	
g		Landé g-factor	
γ	$-\frac{e}{2m_e}g$	gyromagnetic ratio[11]	C kg^{-1}
α	$\frac{d\boldsymbol{M}}{dt} = \ldots + \frac{\alpha}{M_S}\boldsymbol{M}\times\frac{d\boldsymbol{M}}{dt}$	damping parameter	
α_0	$\alpha = \alpha_0 + \alpha_p$	intrinsic damping parameter	
α_p		extrinsic damping parameter	
Heat transport			
T		temperature	K
α	$\boldsymbol{J}_e = -\bar{\bar{\alpha}}\cdot\nabla T$	thermoelectric conductivity	S m^{-1} V K^{-1}
S_{ii}	$\alpha_{ii} = \sigma_{ii}S_{ii}$	Seebeck coefficients	V K^{-1}
Skyrmion transport			
Q	$\frac{1}{4\pi}\iint\widehat{\boldsymbol{M}}\cdot\left(\partial_x\widehat{\boldsymbol{M}}\times\partial_y\widehat{\boldsymbol{M}}\right)dxdy$	topological index	
$\boldsymbol{G}$	$-\frac{M_S d_F}{\lvert\gamma\rvert}4\pi Q\widehat{z}$	gyromagnetic coupling vector	kg s^{-1}
D_{ij}	$\frac{M_S d_F}{\lvert\gamma\rvert}\iint\left(\partial_i\widehat{\boldsymbol{M}}\cdot\partial_j\widehat{\boldsymbol{M}}\right)didj$	dyadic coefficients	kg s^{-1}

(1)In the literature, the terms moment $\boldsymbol{m}$ and spin $\boldsymbol{s}$ are sometimes used interchangeably, and the ↑ and ↓ symbols sometimes refer to $\boldsymbol{m}$ or $\boldsymbol{s}$, which can lead to confusion as $\boldsymbol{m}$ and $\boldsymbol{s}$ are of opposite sign from $\boldsymbol{s} = -\boldsymbol{m}/|\gamma|$. In this book, the term moment refers exclusively to $\boldsymbol{m}$ and spin to $\boldsymbol{s}$; the ↑ and ↓ symbols refer exclusively to the orientation of $\boldsymbol{m}$, and the ⇑ and ⇓ symbols refer exclusively to the orientation of $\boldsymbol{s}$. Note that, in the literature, the + and − symbols are sometimes used for the orientation of $\boldsymbol{s}$.

(2)In the literature, the spin accumulation is defined either as $\boldsymbol{\mu}_s = \boldsymbol{\mu}^{\uparrow} - \boldsymbol{\mu}^{\downarrow}$, or $\boldsymbol{\mu}_s = -(\boldsymbol{\mu}^{\uparrow} - \boldsymbol{\mu}^{\downarrow})$, or $\boldsymbol{\mu}_s = \pm(\boldsymbol{\mu}^{\uparrow} - \boldsymbol{\mu}^{\downarrow})/2$, resulting in a factor $\pm 1/2$ difference in some of the equations depending on the articles and books consulted. Throughout this textbook, the term spin refers to $\boldsymbol{s}$ and the spin $\boldsymbol{s}$ accumulation is thus defined as $\boldsymbol{\mu}_s = \boldsymbol{\mu}^{\Uparrow} - \boldsymbol{\mu}^{\Downarrow} = -(\boldsymbol{\mu}^{\uparrow} - \boldsymbol{\mu}^{\downarrow})$, because the ↑ and ↓ symbols refer to the orientation of the moment $\boldsymbol{m}$, whereas the ⇑ and ⇓ symbols refer to the orientation of the spin $\boldsymbol{s}$, which points in the opposite direction to $\boldsymbol{m}$.

(3)In the literature, l_{sf}^* is most often written l_{sf}. The notation l_{sf} may implicitly infer that the spin-diffusion length is spin-independent, which is not always the case. Hence, we prefer to use a superscript as a warning.

(4)It is customary to incorrectly use the term spin conductance for σ/l_{sf}^*, although its unit is S.m^{-2}. Note that this is the same unit as $G^{\uparrow\downarrow}/A$. In practice, the spin conductance is often used per quantum conductance $\frac{\sigma/l_{sf}^*}{G_0}$. It can be compared directly to $g^{\uparrow\downarrow} = \frac{G^{\uparrow\downarrow}}{AG_0/2}$.

(5)In the literature, the spin mixing conductance is defined either as $G^{\uparrow\downarrow}$, in units of S as here and throughout this textbook, or as $G^{\uparrow\downarrow}/A$, in which case it is in units of S.m^{-2}, resulting in a factor A difference in some of the equations depending on the articles and books consulted.

(6)The coefficients $t_{\uparrow(\downarrow)}$ and $r_{\uparrow(\downarrow)}$ account for the transmission and reflection of a moment $\boldsymbol{m}$ by a ferromagnet of magnetization $\boldsymbol{M}$. Equivalently, the coefficients $t_{\Uparrow(\Downarrow)}$ and $r_{\Uparrow(\Downarrow)}$ account for transmission and reflection of a spin $\boldsymbol{s} = -\frac{1}{|\gamma|}\boldsymbol{m}$ by a ferromagnet of macrospin density $\boldsymbol{S}_M = -\frac{1}{|\gamma|}\boldsymbol{M}$. Therefore, $t_{\Uparrow(\Downarrow)} = t_{\uparrow(\downarrow)}$ and $r_{\Uparrow(\Downarrow)} = r_{\uparrow(\downarrow)}$, $G^{\uparrow\downarrow} = \frac{e^2}{h}\sum_{n\in N}\left[1 - \sum_{m\in N} r_{\uparrow}^{nm}\left(r_{\downarrow}^{nm}\right)^*\right] = \frac{e^2}{h}\sum_{n\in N}\left[1 - \sum_{m\in N} r_{\Uparrow}^{nm}\left(r_{\Downarrow}^{nm}\right)^*\right]$, and $g^{\uparrow\downarrow} = 1 - \left(t_{\uparrow}t_{\downarrow}^* + r_{\uparrow}r_{\downarrow}^*\right) = 1 - \left(t_{\Uparrow}t_{\Downarrow}^* + r_{\Uparrow}r_{\Downarrow}^*\right)$.

(7)It is customary to incorrectly use the term spin mixing conductance for $g^{\uparrow\downarrow} = \frac{G^{\uparrow\downarrow}}{AG_0/2}$, although its unit is m^{-2}.

(8)In the literature, the spin current density $\boldsymbol{J}_s = -(\boldsymbol{j}^{\uparrow} - \boldsymbol{j}^{\downarrow})$ is expressed either in A.m^{-2}, units of charge current density, as here and throughout this textbook, or in J.m^{-2}, units of the density of spin angular momentum $\frac{\hbar}{2}$, in which case $\boldsymbol{J}_s = -\frac{\hbar}{2e}(\boldsymbol{j}^{\uparrow} - \boldsymbol{j}^{\downarrow})$, resulting in a factor $\frac{\hbar}{2e}$ difference is some of the equations depending on the articles and books consulted. In this textbook, we use $\boldsymbol{Q} = \frac{\hbar}{2e}\boldsymbol{J}_s$. In the literature, the spin current is also defined either as $\boldsymbol{J}_s = \boldsymbol{j}^{\uparrow} - \boldsymbol{j}^{\downarrow}$, or as $\boldsymbol{J}_s = -(\boldsymbol{j}^{\uparrow} - \boldsymbol{j}^{\downarrow})$, resulting in a factor −1 difference in some of the equations depending on the articles and books consulted. Throughout this textbook, the term spin refers to $\boldsymbol{s}$ and the spin $\boldsymbol{s}$ current is thus defined as $\boldsymbol{J}_s = \boldsymbol{j}^{\Uparrow} - \boldsymbol{j}^{\Downarrow} = -(\boldsymbol{j}^{\uparrow} - \boldsymbol{j}^{\downarrow})$, because the $\uparrow$ and $\downarrow$ symbols refer to the orientation of the moment $\boldsymbol{m}$, whereas the $\Uparrow$ and $\Downarrow$ symbols refer to the orientation of the spin $\boldsymbol{s}$, which points in the opposite direction to $\boldsymbol{m}$.

(9)In the literature, $\boldsymbol{m}$ is used either to define a property, in A.m^{-1}, units of magnetization, as throughout this textbook, or for the dimensionless magnetization unit vector, in which case $\boldsymbol{m} = \frac{\boldsymbol{M}}{M_S}$, resulting in some possible misunderstanding and a factor M_S difference is some of the equations depending on the articles and books consulted. In this textbook, we use an *overcaret* for unit vectors, *i.e.*, $\widehat{\boldsymbol{M}} = \frac{\boldsymbol{M}}{M_S}$.

(10)In the literature, the torque $\boldsymbol{T} \equiv -\frac{1}{|\gamma|}\frac{d\boldsymbol{M}}{dt}$ is expressed either in J.m^{-3}, as here and throughout this textbook, or in A.m^{-1}.s^{-1}, in which case $\boldsymbol{T} \equiv -\frac{d\boldsymbol{M}}{dt}$, resulting in a factor $|\gamma|$ difference in some of the equations depending on the articles and books consulted.

(11)In the literature, the gyromagnetic ratio $\gamma = -\frac{e}{2m_e}g$, which is negative for electron spin is used either as is in the equations, *e.g.* $\frac{d\boldsymbol{M}}{dt} = \gamma\boldsymbol{M} \times (\mu_0\boldsymbol{H}_{\text{eff}})\ldots$, or in absolute value $|\gamma|$, as throughout this textbook, *e.g.* $\frac{d\boldsymbol{M}}{dt} = -|\gamma|\boldsymbol{M} \times (\mu_0\boldsymbol{H}_{\text{eff}})\ldots$, or replaced by $\gamma_0 = -\gamma\mu_0$, to make it positive and include μ_0 to fit the cgs unit system: $\frac{d\boldsymbol{M}}{dt} = -\gamma_0\boldsymbol{M} \times \boldsymbol{H}_{\text{eff}}\ldots$, resulting in a factor -1 or $\pm\mu_0$ difference is some of the equations depending on the articles and books consulted.

sepec numérique
Impression & brochage - France
Numéro d'impression : N04545230313 - Achevé d'imprimer : Avril 2023
Dépôt légal : Avril 2023
IMPRIM'VERT®
PEFC 10-31-3532 / Certifié PEFC / Ce produit est issu de forêts gérées durablement et de sources contrôlées. / pefc-france.org